Springer Series in Reliability Engineering

Series Editor

Hoang Pham

For further volumes:
http://www.springer.com/series/6917

Gilberto Francisco Martha de Souza
Editor

Thermal Power Plant Performance Analysis

Springer

Gilberto Francisco Martha de Souza
Department of Mechatronics and Mechanical Systems
Polytechnic School
University of São Paulo
05508-900 São Paulo
Brazil
e-mail: gfmsouza@usp.br

ISSN 1614-7839 e-ISSN 2191-5377
ISBN 978-1-4471-2308-8 e-ISBN 978-1-4471-2309-5
DOI 10.1007/978-1-4471-2309-5
Springer London Dordrecht Heidelberg New York

British Library Cataloguing in Publication Data
A catalogue record for this book is available from the British Library

Library of Congress Control Number: 2011943798

Printed on acid-free paper

Springer is part of Springer Science+Business Media (www.springer.com)

To Leonardo, Livia, Cleusa and Maria da Conceição

Contents

Introduction

Gilberto Francisco Martha de Souza

Abstract This chapter presents the motivation for the development of the present book. Because the need for electricity is pervasive in our society, there is a continuing interest in the technology of electric power production and distribution. The chapter presents some forecasts of electric power production that indicate the massive use of thermal power plants, fired with coal or natural gas. In order to improve the efficiency of those power plants, the use of Overall Equipment Effectiveness (OEE) as a key performance indicator is discussed. Finally the link between reliability and maintainability concepts and the OEE index is presented.

1 Introduction

According to the report International Energy Outlook (IEO) 2010 [2] the world net electricity generation projection increases by 87%, from 18.8 trillion kilowatt hours in 2007 to 25.0 trillion kilowatt hours in 2020 and 35.2 trillion kilowatt hours in 2035. Although the recession slowed the rate of growth in electricity demand in 2008 and 2009, growth returns to pre-recession rates by 2015. In general, in OECD countries, where electricity markets are well established and consumption patterns are mature, the growth of electricity demand is slower than in non-OECD countries, where a large amount of potential demand remains unmet. According to that report, the total net generation in non-OECD countries increases by 3.3% per year on average, as compared with 1.1% per year in OECD nations.

G. F. M. de Souza (✉)
Department of Mechatronics and Mechanical Systems Engineering,
Polytechnic School, University of São Paulo, Av. Prof. Mello Moraes, 2231,
05508-900 São Paulo, Brazil
e-mail: gfmsouza@usp.br

G. F. M. de Souza (ed.), *Thermal Power Plant Performance Analysis*,
Springer Series in Reliability Engineering, DOI: 10.1007/978-1-4471-2309-5_1,

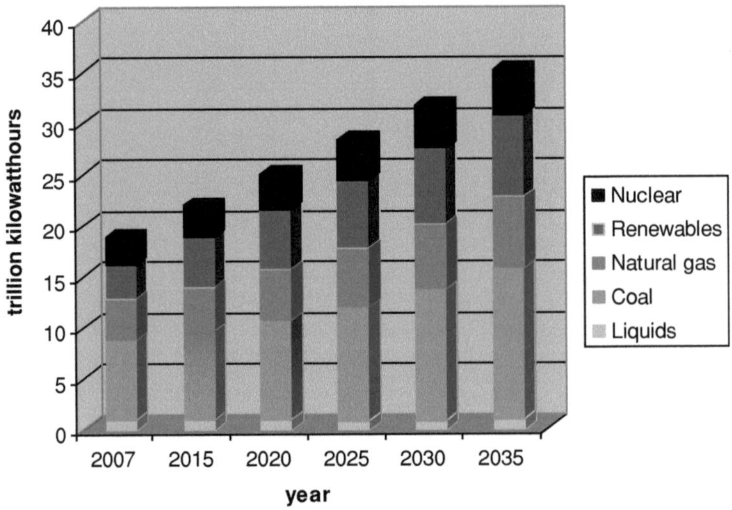

Fig. 1 Forecast of world net electricity generation by fuel, 2007–2030, DoE [2]

The OECD (Organization for Economic Co-operation and Development) provides a forum in which governments can work together to share experiences and seek solutions to common problems. The following countries are consider as OCDE members for the statistics of International Energy Outlook: the United States, Canada, Mexico, Austria, Belgium, Czech Republic, Denmark, Finland, France, Germany, Greece, Hungary, Iceland, Ireland, Italy, Luxembourg, the Netherlands, Norway, Poland, Portugal, Slovakia, Spain, Sweden, Switzerland, Turkey, the United Kingdom, Japan, South Korea, Australia, and New Zealand.

The rapid increase in world energy prices from 2003 to 2008, combined with concerns about the environmental consequences of greenhouse gas emissions, has led to renewed interest in alternatives to fossil fuels—particularly, nuclear power and renewable resources. As a result, long-term prospects continue to improve for generation from both nuclear and renewable energy sources—supported by government incentives and by higher fossil fuel prices.

According to DoE [2] from 2007 to 2035, world renewable energy use for electricity generation will grow by an average of 3.0% per year, as shown in Fig. 1, and the renewable share of world electricity generation will increase from 18% in 2007 to 23% in 2035. Coal-fired generation increase forecast is an annual average of 2.3%, making coal the second fastest-growing source for electricity generation in the projection. The outlook for coal could be altered substantially, however, by any future legislation that would reduce or limit the growth of greenhouse gas emissions. Generation from natural gas and nuclear power—which produce relatively low levels of greenhouse gas emissions (natural gas) or none (nuclear)—according to projections, will increase by 2.1 and 2.0% per year, respectively.

The category liquids include petroleum based fuels, such as Diesel oil or crude oil, and the category renewable includes hydroelectric, wind and other renewable electric power generation.

Those projections are based on a business-as-usual trend estimate, given known technology and technological and demographic trends. The IEO 2010 cases assume that current laws and regulations are maintained throughout the projections.

Most of the world's electricity is produced at thermal power plants (TPP), which use traditional fuels, coal, gas and fuel oil, and up to 20% of the world's electricity is produced by hydroelectric power plants (HPP). In countries with well-to hydropower, the figure is much higher: Norway (99%), Brazil (92%), Austria, Canada, Peru, New Zealand—over 50%.

According to the DoE [2] forecast, coal-fired generation accounted for 42% of the world electricity supply. Sustained high prices for oil and natural gas make coal-fired generation more attractive economically.

The natural gas as an energy source for electric power generation is attractive for combined-cycle power plants because of its fuel efficiency and relative low emissions.

The coal-fired power plants are based on Rankine thermodynamic cycle. This facility generates electricity by producing steam in a steam generator and expanding the steam through a turbine coupled to an electrical generator. The same Rankine cycle can be used with liquid fuels.

The natural gas fired plants are based on Brayton thermodynamic cycle with combustion turbines, in either simple or combined-cycle applications. Those combustion turbines can also be adapted to operate as dual fuel machines, using Diesel oil or natural gas as fuel.

The thermal power plants are very important for social development and must be designed and operated according to the most suitable available technologies. The final product, the electrical generation, must reflect responsible application of economic and engineering principles based on social and environmental concerns.

The purpose of this book is to discuss the operational aspects associated with thermal power plants aiming at not only achieve thermodynamic based performance standards but also performance index associated with environmental and operational aspects.

2 Performance Index

Process companies are adopting a new consolidated approach to performance improvement based upon the use of a KPI (key performance indicator) known as Overall Equipment Effectiveness (OEE).

OEE is a very simple metric to immediately indicate the current status of a industrial process and also a complex tool allowing you to understand the effect of the various issues in the process and how they affect the entire process. OEE can be calculated as:

$$\text{OEE} = \text{Availability} \times \text{Performance} \times \text{Quality} \tag{1}$$

Availability refers to the process equipment being available for production when scheduled. At the most basic level, when a process is running it is creating value for the end user. When a process is stopped, it's creating a cost with no associated value. Whether it's due to mechanical failure, raw materials or operator issues, a piece of equipment is either producing or not producing. By comparing scheduled run time to actual run time, the availability component of OEE allows for a determination of lost production due to down time.

Performance is determined by how much waste is created through running at less than optimal speed. Performance allows for a determination of how much production was lost by cycles that did not meet the ideal cycle time.

Quality focuses on identifying time that was wasted by producing a product that does not meet quality standards. By comparing the quantity of good to reject parts the percent of time actually adding value by producing good product is exposed.

The definition used for general industrial process can be adapted for electricity generation.

The performance of the thermal power plant can be represented by its efficiency. The efficiency of a power plant is usually measured as a ratio of its electrical output to the amount of heat used, expressed as a percentage. Typical commercial plants range from about 30–65% efficiency. The more efficient plants cost more to build. Efficiency depends more on how the energy is used rather than how it is produced because ratings are based on conversion of heat to electrical power.

The quality of the power plant is associated with the parameters (voltage and frequency) of the generated electricity in comparison with the required standards.

The availability of the power plant is associated with the reliability and maintenance planning of each piece of equipment installed in the plant. The availability depends also on the skills of operators and maintenance teams.

The use of OEE index can help the electricity-generating power-stations managers to investigate theirs competence in maintaining reliable equipment at competitive costs.

Although that index can be used to evaluate changes in operational procedures or equipment update aiming at improving plant performance the plant managers use it to highlight the strengths and weakness of equipment maintenance practices.

According to Eti, Ogaji and Probert [3] the availability and quality rate for the world's best power- stations are higher than 98%. The OEE can also be used as index to demonstrate the relation between the plant performance and the issues recommended by PAS 55 [1].

PAS 55 is the British Standards Institution's (BSI) Publicly Available Specification for the optimized management of physical assets. The specification states that organizations must establish, document, implement, maintain, and continually improve their asset management system. In this context, asset management system refers collectively to the overall policy, strategies, governance, plans and actions of an organization regarding its asset infrastructure.

Aiming at discussing the aspects associated with performance evaluation of thermal power plants, the book presents chapters associated with thermal and environmental performance of power plants and also presents the concepts of reliability, maintenance and risk analysis applied to power plant management. Each chapter is written by an expert in the subject.

3 Chapter Contents

After a brief introduction to the book, in Chap. 2 it is reviewed the fundamental principles of Thermodynamics aiming at its application to power plants cycle analysis. Next, the three most common thermodynamic cycles are studied starting with the Brayton cycle, the Diesel Cycle, and the Rankine cycle. These ideal cycles are thermodynamic operating models for gas turbines, diesel engines, and steam turbines, respectively. Thermal efficiencies, operating conditions and cycle variations are also analyzed.

Chapter 3 presents the typical thermal power plants configuration exploring the equipment technology used for each configuration. The chapter discusses gas and steam turbines, steam generators (including heat recovery steam generators) and heat exchangers. It also discusses the efficiency and operational aspects of each plant configuration in single or combined-cycle.

Chapter 4 presents the environmental impacts of the thermoelectric plants installation and operation. The most significant impacts occur during operation because solid, liquid and gaseous wastes are generated continuously and permanently in significant quantities. The magnitude of the impacts depends mainly on the amount, type of waste and the ability of the environment to absorb them. Their nature, quantities and chemical and physical characteristics depended mainly on both the technology and the fuel employed in the power plant.

In Chap. 5 the basic definitions supporting component and system reliability analysis are presented. Reliability and failure rate curves are presented aiming at providing information regarding the failure modes of components. The Failure Mode and Effects Analysis (FMEA) and Fault Tree Analysis (FTA) methods for system reliability analysis are presented.

In Chap. 6 the basic concepts associated to maintenance planning are presented. In order to improve power plant maintenance planning the Reliability Centered Maintenance philosophy is detailed presented. The improvement in maintenance planning has a direct effect on power plant availability, increasing the plant OEE.

In Chap. 7 the basic concepts associated with risk analysis of complex systems are presented, including the discussion of risk quantification in a systems framework. Risk-informed decision making is introduced on the basis of benefit-to-cost analysis. Those concepts can be applied to decision making problems related to power plant design and operational profile changes.

Chapter 8 presents the application of reliability concepts to evaluate the overall performance of gas turbine used in open cycle or combined-cycle thermal power plants. The thermodynamics derived performance parameters of gas turbines are presented, including the presentation of the tests codes used to evaluate turbine performance during power plant commissioning. The reliability and availability concepts associated with gas turbine are presented and an example of reliability analysis of a heavy duty gas turbine is also presented.

In Chap. 9 the reliability and maintainability concepts are used to evaluate a combined-cycle power plant. The most critical components as for power plant reliable performance are identified. Based on the plant operational profile the reliability and availability are estimated. A more detailed analysis of the cooling tower system is executed once the failures of that system not only affect the plant nominal output but strongly affect plant availability.

Finally, Chap. 10 presents the basic concepts associated with Risk-based Inspection and Maintenance (RBIM) philosophy and their application in maintenance planning aiming at controlling power plant equipment degradation. The method is customized for power plant analysis considering the constraints associated with that application.

References

1. British Standards Institution (BSI) (2008) PAS (Publicly Available Specifications) 55: Asset Management—Part 1 & 2. BSI, London
2. Department of Energy (DoE) (2010) EIA-0484 international energy outlook 2010. U.S. Department of Energy, Washington
3. Eti M, Ogaji S, Probert S (2005) Maintenance schemes and their implementation for the Afam thermal-power station. Appl Energy 82:255–265

Fundamentals of Thermodynamics Applied to Thermal Power Plants

José R. Simões-Moreira

Abstract In this chapter it is reviewed the fundamental principles of Thermodynamics aiming at its application to power plants cycle analysis. The three most common thermodynamic cycles are studied starting with the Brayton cycle, the Diesel Cycle, and the Rankine cycle. These ideal cycles are thermodynamic operating models for gas turbines, diesel engines, and steam turbines, respectively. Thermal efficiencies, operating conditions and cycle variations are also analyzed. The last issue studied is the combined Brayton-Rankine cycle, which is a trend in industry due to its higher overall efficiency.

1 Thermodynamics Principles

In this section is presented a review of fundamental thermodynamic principles, thermodynamic properties, and the governing laws applied to processes commonly presented in thermal machines.

1.1 Thermodynamic Properties, Equations and Tables

Specific internal energy, u—is the energy stored in the substance due to molecular motion as well as intermolecular forces. The SI unit is kJ/kg.

J. R. Simões-Moreira (✉)
Mechanical Engineering Department at Escola Politécnica da USP,
SISEA Alternative Energy Systems Lab, São Paulo, Brazil
e-mail: jrsimoes@usp.br

G. F. M. de Souza (ed.), *Thermal Power Plant Performance Analysis*,
Springer Series in Reliability Engineering, DOI: 10.1007/978-1-4471-2309-5_2,

Specific enthalpy, h—is the sum of the specific internal energy and the product of pressure P versus specific volume, v. The SI unit is kJ/kg.

$$h = u + Pv \tag{1}$$

Kinetic energy, KE—is the energy a system of mass m possesses due to the macro fluid motion at velocity V.

$$KE = mV^2/2 \tag{2}$$

Potential energy, PE—is the energy due to the gravitational field g that a mass m possess in relation to a height z from a reference plane.

$$PE = mgz \tag{3}$$

Shaft work, W—is the mechanical work produced or absorbed by the rotating shaft of the thermal machine.

Shaft power, $\dot{W}$—is the mechanical power produced or absorbed by the rotating shaft of the thermal machine.

Heat, Q—is the form of energy transferred to or from the machine due to a difference of temperatures between the machine and the surroundings, the higher temperature to the lower one.

Thermal power, $\dot{Q}$—is the form of energy rate transferred to or from the machine due to a difference of temperatures between the machine and the surroundings, the higher temperature to the lower one.

Phase change: pure substances have molecular arrangement in phases. A solid phase is the one in which the molecules do not move freely, such as in ice. In liquid phase, the molecules move partially free, such as in liquid water. Finally, in vapor phase the molecules move freely, such as in steam. All pure substances have those three phases. It is also possible to have different solid phases.

Figure 1 shows a phase diagram for water in the temperature x specific volume plane for the liquid–vapor phases. The "bell shape" curve is more appropriately known as the saturation curve. The liquid phase is on the left and the vapor phase is on the right region. Inside the "bell shape" is the two-phase region, where liquid and vapor phases coexist in thermodynamic equilibrium. The left line is known as saturated liquid and the right one is the saturated vapor. The saturation lines meet at the critical point. All states to the left of the saturation liquid line is compressed liquid and the states to the right of the saturation vapor line are superheated vapor.

Substances change states. Consider compressed liquid water at, say, room temperature and normal pressure indicated by state 1 in the piston-cylinder setup on the right of Fig. 1. As heat is supplied at constant pressure, the system temperature increases until the liquid saturation line is achieved at state 2. If heat continues to be supplied a liquid–vapor phase change takes place and vapor bubbles arise until all the liquid phase undergoes a vaporization process and only vapor is seen inside the piston-cylinder device at state 3, or saturated vapor. On continuing supplying heat the saturated vapor becomes superheated vapor, state 4.

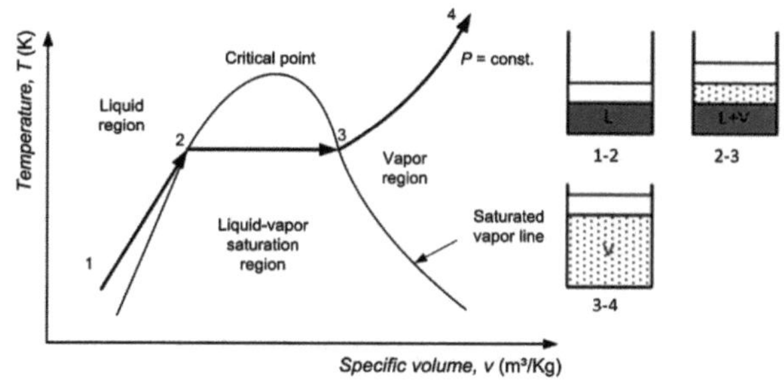

Fig. 1 Liquid-vapor saturation curve in the temperature-specific volume plane and an illustration of a liquid-vaporphase change process at constant pressure

Of course, if one starts as a superheated vapor (state 4) a liquid state 1 can also be attained by removing heat from the system. If the experiment is carried out at a higher pressure, the same behavior will be observed, except that the phase change will start at a higher temperature.

There is a direct correspondence between pressure and temperature during a phase change process, which is known as the saturation curve. For each substance, including water, there is a specific temperature where a phase change will occur at a given pressure. Conversely, there is a specific pressure where a phase change will occur at a given temperature. However, for pressure above the critical pressure, there will be no phase change, as the two saturation lines meet as at the critical point as seen in Fig. 1. Therefore, above the critical pressure and temperature there will be no liquid–vapor phase change.

The process illustrated in Fig. 1 takes place at a constant pressure, known as isobaric, which is imposed on the system by the piston weight plus local atmospheric pressure. Other relevant thermodynamic processes are: (a) isothermal—constant temperature; (b) isochoric—constant specific volume; (c) adiabatic—no heat transfer to or from the system; (d) reversible process—no "losses" in the process. Of course, these processes are general and they can occur with or without any phase change.

Precise thermodynamic properties of water and many other substances can be found in tables presented in basic thermodynamic books. Normally, there are two sets of tables for water. One is valid only for the liquid–vapor saturation region, and the other for the superheated vapor region. The saturation table provides saturation liquid and vapor properties, while the other table provides superheated vapor properties.

Vapor quality, x—is defined as the ratio between the vapor mass, m_v, and the total mass, m_T, in a given system. Vapor quality is a thermodynamic property valid only for the two-phase region or saturation region, where a mixture of liquid and vapor are at thermodynamic equilibrium.

$$x = \frac{m_v}{m_T} \tag{4}$$

Thermodynamic properties such as specific volume, specific internal energy, and specific enthalpy are averaged by the vapor quality in the two-phase region from the saturated liquid (subscript "L") and vapor (subscript "V") corresponding values. Average saturation properties can be obtained from a saturation table such as the one for water.

$$v = xv_v + (1 - x)v_L \tag{5}$$

$$u = xu_v + (1 - x)u_L \tag{6}$$

$$h = xh_v + (1 - x)h_L \tag{7}$$

Equations of State and Specific Heats thermodynamic properties are related to each other by equations of state. Most equations of state relate pressure, specific volume, and temperature, and have the general form given by $f(P, v, T) = 0$. An equation of state, or simply, EOS can be a very complex mathematical function having several coefficients and constants and can be valid for both liquid and vapor regions. Also, equations of state can be presented in graphical form and tables. Saturation and superheated tables are good examples of precise equations of state. However, all equations of state valid for the vapor phase do have a low pressure limit given by the ideal equation of state given by

$$Pv = RT \tag{8}$$

where the temperature must be in absolute value, and R is the particular gas constant, which is given by the ratio between the universal ideal gas constant, $\Re$, and the gas molecular weight, M.

$$R = \frac{\Re}{M} \tag{9}$$

Some values of $\Re$ are 8.314 kJ/kgmol.K = 1.987 kcal/kgmol.K = 847.7 kgf/kgmol.K.

All vapors and gases agree with the ideal EOS for pressures much lower than the critical pressure and the ideal EOS can be used for system pressure lower than 5% for engineering purposes. The lower the pressure, the better the agreement. Also, if the system temperature is about twice the critical temperature, the ideal behavior is valid as well.

Specific heat at constant pressure, C_P strictly speaking this thermodynamic property is defined in terms of partial derivative. However, when the substance is an ideal gas, it can defined as

$$C_p = \frac{dh}{dT} \tag{10}$$

Specific heat at constant volume, C_V strictly speaking this thermodynamic property is defined in terms of partial derivative. However, when the substance is an ideal gas, it can defined as

$$C_V = \frac{du}{dT} \tag{11}$$

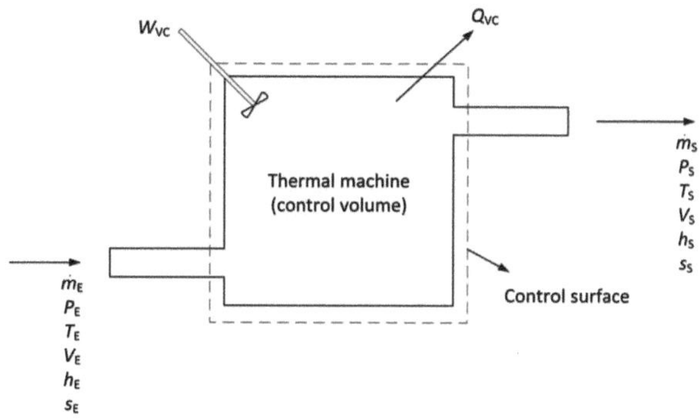

Fig. 2 Control volume for energy balance analysis

The first specific heat involves specific enthalpy, and the other one, specific internal energy. For liquids and solids, both specific heats are very close to each other and one can say simply *specific heat, C.*

For an ideal gas, there is a very useful relationship between these two specific heats given by

$$C_p - C_V = R \tag{12}$$

The ratio between the two specific heats is quite important in analyzing thermodynamic process. For this it is defined the specific heat ratio, γ, as:

$$\gamma = \frac{C_p}{C_V} \tag{13}$$

The above property is very useful on studying some processes in ideal gas. γ is bound by two limits: $1 \leq \gamma \leq 5/3$. For air $\gamma = 1.4$.

1.2 First Law of Thermodynamics Analysis for Control Volumes

Thermal machines convert chemical energy in shaft work by burning fuel (heat) in a combustion chamber. In doing so, mas fluxes of air and fuel enter the machine and combustion products exit it. In a working machine, energy in its several forms is presented in the conversion process, such as heat, shaft work, enthalpy, and chemical energy. Even though energy is transformed from one form into another, the overall amount of energy must be conserved as stated by the First Law of Thermodynamics or Law of Conservation of Energy. In order to establish the First Law consider the schematics in Fig. 2 showing a control volume around a thermal machine. All relevant forms of energy and variables fluxes are shown along with shaft power and heat flux.

Energy balance for the control volume in Fig. 2 results in

$$\left(\frac{dE}{dt}\right)_{CV} = \sum \dot{m}_i \left(h_i + \frac{V_i^2}{2} + Z_i\right) - \sum \dot{m}_o \left(h_o + \frac{V_o^2}{2} + Z_o\right) + \dot{Q} - \dot{W}. \quad (14)$$

The total energy E is the instantaneous total energy within the control volume (such as a thermal machine). The first two terms in the r.h.s. are the specific enthalpy and the kinetic and potential specific energies associated with all inlet mass fluxes, $\dot{m}_i$, and outlet mass fluxes, $\dot{m}_o$. $\dot{Q}$ is the rate of heat the control volume exchange with the environment, and $\dot{W}$is the shaft power generated by the control volume. The units of shaft power and rate of heat transfer is kW in the SI system. Positive values are for heat gained and for net work produced.

Most machines operate in steady state. In steady states, the heat rate and shaft power along with the inlet and the outlet conditions and thermodynamic properties do not change and, consequently, the total energy do not vary in time. Therefore, the time rate is null and the Eq. 14 can be simplified to obtain

$$\sum \dot{m}_i \left(h_i + \frac{V_i^2}{2} + Z_i\right) + \dot{Q} = \sum \dot{m}_o \left(h_o + \frac{V_o^2}{2} + Z_o\right) + \dot{W}. \quad (15)$$

A particular case arises when there is only one inlet and one outlet mass flux. So,

$$h_i + \frac{V_i^2}{2} + Z_i + q = h_o + \frac{V_o^2}{2} + Z_o + w. \quad (16)$$

where lower cases are used for heat and work terms on mass basis, i.e., heat per unit of mass and work by unit of mass. Also, inlet and outlet mass fluxes are equal.

1.3 Second Law of Thermodynamics Analysis for Control Volumes

The rate of entropy generated in a control volume (Fig. 2) can be written according to Eq. 17

$$\left(\frac{dS}{dt}\right)_{CV} \geq \sum \dot{m}_i s_i - \sum \dot{m}_o s_o + \sum \frac{\dot{Q}_{CV}}{T}. \quad (17)$$

where, S is the total instantaneous entropy of the control volume, s_i and s_o are the specific entropy associated with the inlet and outlet mass fluxes, T is the control volume surface temperature where heat is exchanged with the surrounding environment at a given rate, $\dot{Q}_{cv}$. The inequality is to take into account the irreversibilities that can occur. By adding a positive term to consider the time rate at which irreversibility occurs, $\dot{S}_{gen}$, one can drop the inequality. Also, in steady state conditions, the control volume instantaneous entropy remains constant. Therefore, with these two assumptions, once can obtain:

$$\sum \dot{m}_o s_o - \sum \dot{m}_i s_i = \sum \frac{\dot{Q}_{CV}}{T} + \dot{S}_{gen} \tag{18a}$$

If there is only one inlet and one outlet, then

$$\dot{m}(s_o - s_i) = \sum \frac{\dot{Q}_{CV}}{T} + \dot{S}_{gen} \tag{18b}$$

For an adiabatic process, there is no heat transfer, therefore

$$\dot{m}(s_o - s_i) = \dot{S}_{gen} \tag{19a}$$

or, considering that the entropy generation time rate is always positive, then

$$s_o \geq s_i \tag{19b}$$

where, the equality is valid for an adiabatic and reversible ($\dot{S}_{gen} = 0$) process.

1.4 Reversible Work, Polytropic Process and Entropy Variation in Ideal Gases

Reversible Work is the shaft work an ideal machine, such as pumps, compressors, turbines, produces or demands on carrying out a given thermodynamic process. There is a differential fundamental thermodynamic relationship derived from the combination of First and Second Laws of Thermodynamics known as the Gibbs equation, given by:

$$du = Tds - Pdv \tag{20a}$$

By differentiating the specific enthalpy (Eq. 1) and introducing the differential form into the above equation, one obtains.

$$dh = Tds + vdP \tag{20b}$$

For a reversible process, the differential form of the specific entropy is

$$ds = \frac{\delta q}{T} \tag{21}$$

On the other hand, the First Law in differential form neglecting the kinetic and potential energy variations can be obtained from Eq. 16, this is

$$\delta q = dh - \delta w \tag{22}$$

Finally, substituting Eqs. 21 and 22 into Eq. 20b after integration, one obtains.

$$w = -\int_{P_i}^{P_o} v dP \tag{23}$$

Equation 23 is a remarkable expression that allows one to calculate the reversible work by unit of mass for any reversible thermodynamic process in steady state in a control volume. The minus signal comes from the convention of positive work produced by the control volume.

Polytropic Process Thermal machines such as internal combustion engines and gas turbines are modeled by air standard cycles, such as Brayton and Diesel cycles discussed later in this chapter. In the modeling process of those thermodynamic cycles, an amount of air undergoes several thermodynamic processes which can be analyzed by using the ideal gas behavior. In doing so, simple working equations arise. Therefore, it is important to analyze the several thermodynamic process associated with an ideal gas transformations. In a broad sense, many useful thermodynamic reversible processes can be analyzed at once by using the concept of polytropic process. Those processes include isothermal, isentropic, isobaric, isochoric, a general process with or without heat transfer as it will be seen. Air standard cycles can also be used to analyze other devices, such as the Ranque-Hilsh or vortex tube, as presented by Simões-Moreira [1].

A general polytropic process is the one that obeys the following relationship between pressure and specific volume

$$Pv^n = const \tag{24}$$

where n is the polytropic coefficient. It can assume any value. Some particular values of n represent a special thermodynamic process, such as:

- Isobaric process (p = constant): $n = 0$;
- Isothermal process (T = constant): $n = 1$;
- Isentropic or adiabatic reversible process (s = constant): $n = \gamma$;
- Isochoric process (v = constant): $n = \infty$

The reversible work by unit of mass can now be calculated for an ideal gas from its definition (Eq. 23) by a process varying from P_1 to P_2

$$w = -\int_{P_1}^{P_2} v dP = -const \int_{P_1}^{P_2} \frac{dP}{P^{1/n}} \tag{25}$$

The constant can be related either to the initial state 1 or the final one 2, as needed or desired, i.e., $P_1 v_1^n = P_2 v_2^n = const$. In order to carry out the integration, it is necessary to separated the integral in two situations, one is for $n = 1$and the other for $n \neq 1$.

If $n = 1$(isothermal process), then from Eq. 25, one obtains:

$$w = -const \int_{P_1}^{P_2} \frac{dP}{P} = -P_1 v_1 \ln\left(\frac{P_2}{P_1}\right) = -P_2 v_2 \ln\left(\frac{P_2}{P_1}\right) = -RT \ln\left(\frac{P_2}{P_1}\right) \quad (26)$$

If $n \neq 1$ (any other polytropic process), then from Eq. 25, one obtains:

$$w = -const \int_{P_1}^{P_2} \frac{dP}{P^{1/n}} = -\frac{n}{n-1}(P_2 v_2 - P_1 v_1) = -\frac{nR}{n-1}(T_2 - T_1) \quad (27)$$

Entropy Variation in Ideal Gases working equations can be obtained for entropy variation in ideal gases. In order to obtain those equations, let us start off with the differential form of the fundamental thermodynamic relationship given by Eq. 20 and substitute both $du = C_v dT$ from Eq. 11 and the ideal gas equation of state from Eq. 8 into it, to obtain

$$ds = C_V \frac{dT}{T} + R \frac{dv}{v} \quad (28)$$

Next, by integrating it between two states of interest, it yields,

$$\Delta s = s_2 - s_1 = C_V \ln\left(\frac{T_2}{T_1}\right) + R \ln\left(\frac{v_2}{v_1}\right) \quad (29)$$

Alternatives forms of the above equations can also be obtained depending on the selecting the two other independent variables out pressure, temperature and specific volume set. They are:

$$\Delta s = s_2 - s_1 = C_p \ln\left(\frac{T_2}{T_1}\right) - R \ln\left(\frac{P_2}{P_1}\right) \quad (30)$$

and

$$\Delta s = s_2 - s_1 = C_V \ln\left(\frac{P_2}{P_1}\right) + C_p \ln\left(\frac{v_2}{v_1}\right) \quad (31)$$

Isentropic process occurs in analysis of ideal machines as we will see in next sections. By equating Eqs. 29–30 to zero, it is possible to obtain corresponding *P*-*v*-*T* isentropic relationships. They are in order:

$$T_1 v_1^{\gamma-1} = T_2 v_2^{\gamma-1}, \quad (32)$$

$$P_1 v_1^{\gamma} = P_2 v_2^{\gamma}, \quad (33)$$

and

$$P_1^{\frac{1-\gamma}{\gamma}} T_1 = P_2^{\frac{1-\gamma}{\gamma}} T_2, \quad (34)$$

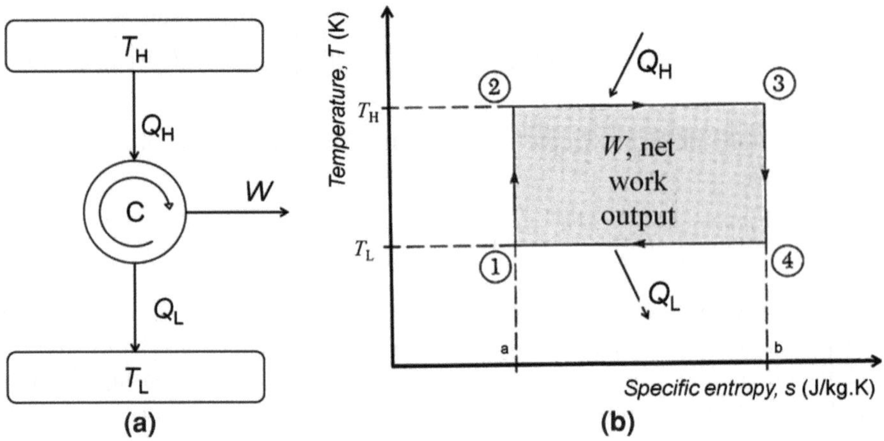

Fig. 3 **a** Schematics of a heat engine; **b** T-s diagram for a Carnot cycle

1.5 The Carnot Cycle

On studying heat engines and thermal machines, one is faced with a question very relevant: Given two sources of thermal energy at two different temperatures, one at a high temperature T_H and the other at a low temperature T_L, what is the maximum conversion of heat drawn from the source at high temperature that can be converted into useful work in an ideal heat engine (reversible one) that operates continuously in a closed thermodynamic cycle? First, the Kelvin-Planck statement of the Second Law of Thermodynamics tells us that it is impossible to have a heat engine that will convert all the heat received from the high temperature source, Q_H, into useful work in a thermodynamic cycle. It is necessary to reject part of the received heat to the low temperature source, Q_L. In other words: it is impossible to have a 100% efficiency heat engine. A schematic of an operating heat engine according to Kelvin-Planck is shown in Fig. 3a.

Second, Carnot devised that the heat engine that can achieve the maximum efficiency in continuously converting heat into work operating between the two heat sources is the one made up of four reversible processes as illustrated in the temperature-entropy diagram in Fig. 3b, which are:

(a) process 1–2—temperature raise from T_L to T_H in an adiabatic reversible process (isentropic);
(b) process 2–3—heat addition, Q_H, in an isothermic reversible process at T_H;
(c) process 3–4—temperature decrease from T_H to T_L in an adiabatic reversible process (isentropic);
(d) process 4–1—heat rejection, Q_L, in an isothermic reversible process at T_L.

The thermal efficiency of any power cycle, η_{th}, is the ratio of the network, W, and the heat received, Q_H.

$$\eta_{th} = \frac{W}{Q_H} = \frac{Q_H - Q_L}{Q_H} = 1 - \frac{Q_L}{Q_H} \tag{35}$$

where, the First Law has also been used, i.e., $W = Q_H - Q_L$.

From the *T-S* in diagram Fig. 3b, it is possible to notice that both heat addition and rejected are associated with entropy variation, i. e.

$$Q_H = T_H \Delta S \tag{36a}$$

and

$$Q_L = T_L \Delta S \tag{36b}$$

Therefore, substituting above equations into Eq. 35, one obtains the final form of the Carnot efficiency, η_C, which is:

$$\eta_C = 1 - \frac{T_L}{T_H} \tag{37}$$

This remarkable result shows that the maximum conversion of heat into work in heat engine operating continuously between two heat sources is limited by the ratio between the two heat sources temperatures. The lower the temperature ratio, the higher the Carnot efficiency. As a final remark, no 100% conversion can take place because it would require either a 0 K low temperature source, or an extremely high temperature source (mathematically, an infinite one), or both.

2 Gas Turbine Cycles

Gas turbines are complex turbo machines made up of thousands of parts. Nevertheless, gas turbines have three main parts that perform the fundamental thermodynamic processes involved in the mechanical shaft power production from the fuel chemical energy as illustrated in Fig. 4. First, the income atmospheric air must undergo a compression process in the compressor section where both pressure and temperature are increased. Next, the compressed air is driven to a combustion chamber where fuel is injected into the compressed air stream and burnt increasing the temperature at a constant pressure process. Finally, the combustion products at a high temperature and pressure are expanded in the power turbine section generating shaft power to drive the compressor as well as an electrical generator or any other rotary device attached to the rotary shaft. The combustion products are exhausted through a nozzle into the atmosphere.

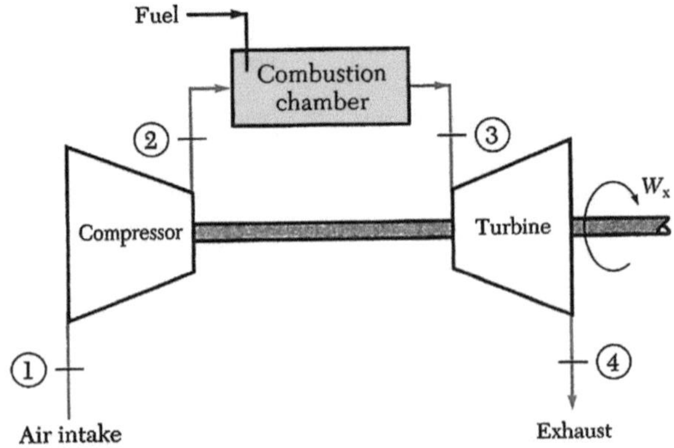

Fig. 4 Three main parts of a gas turbine: the compressor, the combustion chamber, and the power turbine

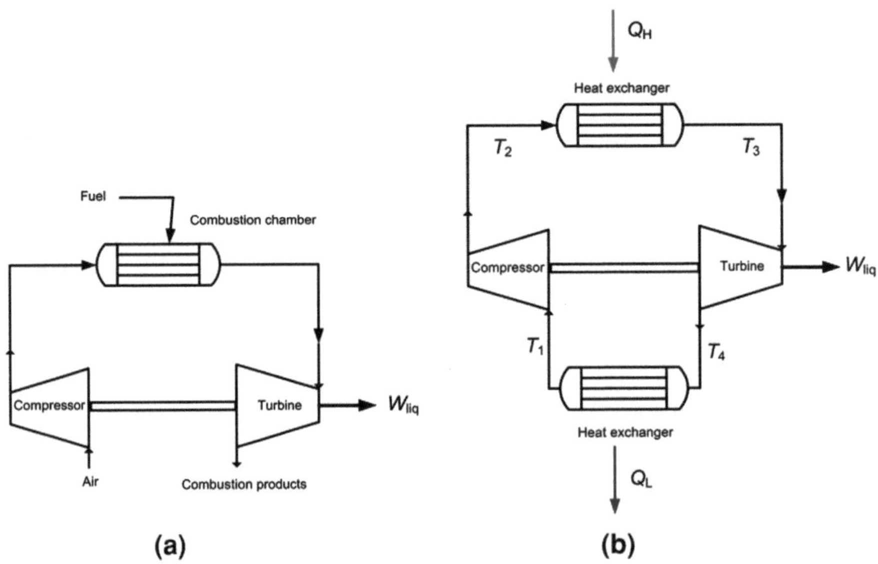

Fig. 5 **a** Open cycle; **b** closed air standard Brayton cycle

2.1 Simple Brayton Cycle

In an actual gas turbine, the working fluid changes from atmospheric air to combustion products that exhaust back to the atmosphere, as illustrated in Fig. 5a. However, in order to evaluate the machine from the thermodynamic point-of-view, some assumptions are needed. Firstly, the working fluid is assumed to be plain air, without any chemical transformation due to the combustion. In doing so, the air–fuel combustion process is replaced by a heat addition process at a constant pressure. Secondly, the exhaust and admission processes are replaced by a heat transfer process to the environment, which makes the air to flow continuously in a closed loop as indicated in Fig. 5b. In the closed cycle, air at environment pressure and temperature is first compressed, next it receives heat Q_H and it is followed by an expansion process in the turbine section to, finally, reject heat Q_L at constant pressure. This is the Air-Standard Brayton Cycle.

Having the cycle of Fig. 5b in mind along with the ideal gas behavior and constant thermodynamic properties one may obtain the working equations from an energy balance (Eq. 16) for each cycle component:

$$heat\ addition : q_H = h_3 - h_2 = C_P(T_3 - T_2) \quad (38)$$

$$heat\ rejection : q_L = h_4 - h_1 = C_P(T_4 - T_1) \quad (39)$$

$$compression\ work : w_{comp} = h_2 - h_1 = C_P(T_2 - T_1) \quad (40)$$

$$turbine\ work : w_{turb} = h_3 - h_4 = C_P(T_3 - T_4) \quad (41)$$

$$cycle\ net\ work : w = w_{turb} - w_{comp} \quad (42)$$

Equations 38 through 42 are on mass basis whose unit is kJ/kg in the international system of units, SI. Also, both the kinetics and potential forms of energy have been neglected.

Figures 6a and 6b gives two important thermodynamic diagrams for cycle analysis. The first one is the temperature-entropy diagram and the second one is the pressure-specific volume diagram. The simple Brayton cycle formed by its four basic ideal gas processes is indicated in both diagrams. The cycle net work is given by the enclosed area shown in figures. First, air is compressed ideally (isentropic) in the compressor (process 1–2) increasing both pressure and temperature at expenses of using compression work (w_{comp}) which is supplied by the turbine itself. Second, heat (q_{H}) is added at constant pressure making up the process 2–3, which heats up the air to the highest cycle temperature, T_3. Next, the high pressure and temperature air undergoes an expansion process (process 3–4) generates work (w_{turb}) enough to drive the compressor and produce net shaft work (w). Finally, heat (q_{L}) is rejected to the environment (process 4–1) at constant low pressure closing the cycle.

The thermal efficiency, η_{th}, of a cycle is defined as the ratio between the cycle net work and heat added, as given by Eq. 35. By applying the First Law for the whole cycle, one easy can show that $w = q_H - q_L$. Therefore, one obtains:

$$\eta_{th} = 1 - \frac{q_L}{q_H} \quad (43)$$

Finally, using Eqs. 38 and 39, it yields:

$$\eta_{th} = 1 - \frac{C_P(T_4 - T_1)}{C_P(T_3 - T_2)} = \frac{T_1(T_4/T_1 - 1)}{T_2(T_3/T_2 - 1)} \quad (44)$$

By examining the temperature-entropy diagram in Fig. 6a, one can easily notice that T_3 is the maximum cycle temperature, also known as the firing temperature, while T_1 is the minimum one (usually the environment temperature).

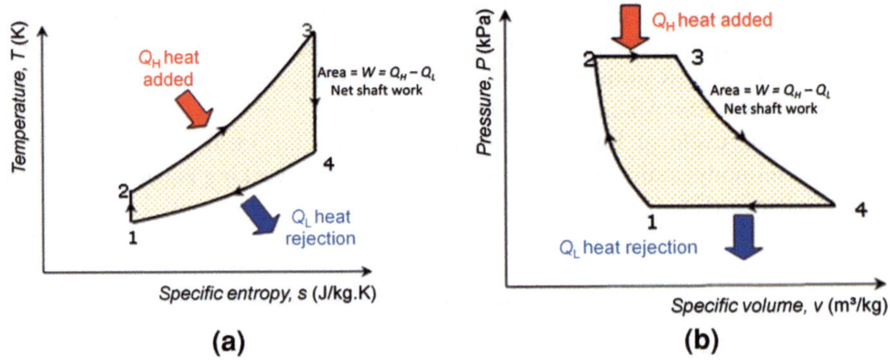

Fig. 6 Simple Brayton cycle in **a** temperature-specific entropy diagram; **b** pressure-specific volume diagram

By using isentropic ideal gas relationships between pressure and temperature (Eq. 34), it is straightforward to show that:

$$\frac{p_2}{p_1} = \left(\frac{T_2}{T_1}\right)^{\frac{\gamma}{\gamma-1}} \text{ and } \frac{p_3}{p_4} = \left(\frac{T_3}{T_4}\right)^{\frac{\gamma}{\gamma-1}} \tag{45}$$

Also, from the diagram of Fig. 6b, one may notice that

$$r = \frac{p_2}{p_1} = \frac{p_3}{p_4} \tag{46}$$

where, r is the ratio of maximum and minimum cycle pressures. Therefore, after substituting it into Eq. 45, one may show that $T_3/T_2 = T_4/T_1$. Finally, substituting that in Eq. 44, one obtains:

$$\eta_{th} = 1 - \frac{T_1}{T_2} = 1 - \frac{1}{\left(\frac{p_2}{p_1}\right)^{\frac{\gamma-1}{\gamma}}} = 1 - \frac{1}{r^{\frac{\gamma-1}{\gamma}}} \tag{47}$$

Figure 7 shows a graphics of the thermal efficiency as a function of the pressure ratio for air as given by Eq. 47 for air (value of $\gamma = 1.4$).

Generally, it is not enough to carry out a simple thermal efficiency analysis to find the best operational condition of a Brayton cycle. A non-negligible amount of work is required to compress the air from the inlet to the maximum cycle pressure and this work must be supplied by the turbine itself. Therefore, one should examine the net work produced by the system compressor-turbine as a whole. In order to achieve that, first subtract Eq. 41 from Eq. 40, to obtain:

$$\text{Net work} : w = w_{turb} - w_{comp} = C_p(T_3 - T_4) - C_p(T_2 - T_1) \tag{48}$$

After a few manipulations using previous equations, the net shaft work is given by:

$$w = C_p T_1 \left[\left(\frac{T_3}{T_1}\right) \times \left(1 - \frac{1}{r^{\frac{\gamma-1}{\gamma}}}\right) - \left(r^{\frac{\gamma-1}{\gamma}} - 1\right)\right] \tag{49}$$

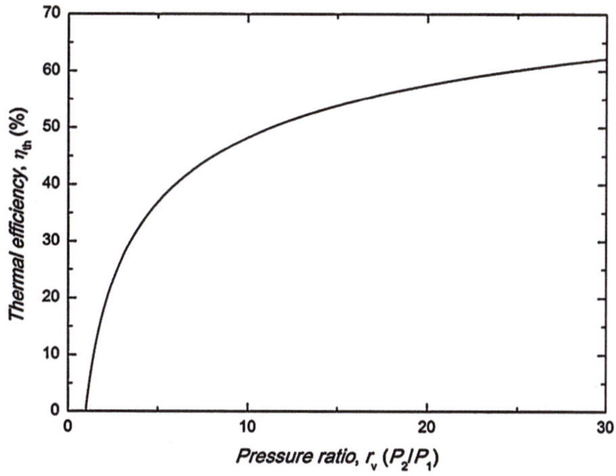

Fig. 7 Thermal efficiency the simple Brayton cycle as a function of the pressure ratio

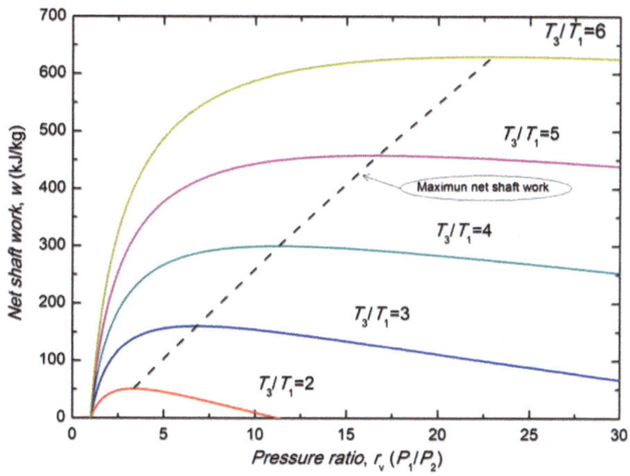

Fig. 8 Net shaft work for several temperature ratios, T3/T1, for inlet air at 300 K

Further analysis of Eq. 49 indicates that the net work is a function of the ratio between maximum and minimum temperature, pressure ratio, along with two fluid thermodynamic properties. It is difficult to get a hold on the precise net shaft work dependency on each one of those variables by a simple straight analysis of that equation, except if one examines it in a parametric graphic form, as shown in Fig. 8. That figure shows the net shaft work for several temperature ratios, i.e., T_3/T_1 assuming an inlet air temperature $T_1 = 300$ K.

The condition of maximum net work is readily obtained by using the simple rule from Calculus, i.e., $\partial w/\partial r)_{T_3/T_1} = 0$.After applying the condition of maximum to Eq. 49 followed by a few manipulations, one obtains the pressure ratio where a maximum net shaft work takes place for a given temperature ratio:

$$r_{\max\ work} = \left(\frac{P_2}{P_1}\right) = \left(\frac{T_3}{T_1}\right)^{\frac{\gamma}{2(\gamma-1)}} \tag{50}$$

In Fig. 8 the condition of maximum net shaft work is indicated by a dashed line.

2.2 Inefficiencies and Actual Brayton Cycle

The actual Brayton cycle is based on real turbo machines that deviate from ideal ones (isentropic). Substantial part of the work produced in the turbine section is drawn by the compressor, which can reach figures as high as 80% of turbine shaft work. If compressor and turbine efficiencies are not high enough, no net shaft work will be generated. Therefore, it is quite important to analyze how much process losses are introduced on the overall performance of the turbine due to machine inefficiency. First, two isentropic definitions must be introduced:

Compressor isentropic efficiency, η_c, is defined as the ratio of ideal or isentropic compression work, w_{comp-a}, to the actual compression work, w_{comp-a}. Figure 9a indicates the ideal and the actual compression process in the T-s diagram.

$$\eta_c = \frac{w_{comp-s}}{w_{comp-a}} = \frac{h_{2s} - h_1}{h_2 - h_1} = \frac{T_{2s} - T_1}{T_2 - T_1} \tag{51}$$

Turbine isentropic efficiency, η_t, is defined as the ratio of the turbine actual work, w_{turb-a}, to the ideal or isentropic turbine work, w_{turb-s}. Figure 9b indicates the ideal and the actual expansion process in the T-s diagram.

$$\eta_t = \frac{w_{turb-a}}{w_{turb-s}} = \frac{h_3 - h_4}{h_3 - h_{4s}} = \frac{T_3 - T_4}{T_3 - T_{4s}} \tag{52}$$

In Fig. 9, one can see both processes of compression (a); both processes of expansion in the turbine section (b); and, finally, on can see the overall combination (c) of those processes.

Using the definition of isentropic compression work (Eq. 51), one can obtain the following equation for the actual compression work.

$$w_{comp-a} = \frac{C_p T_1}{\eta_c}\left(r^{\frac{\gamma-1}{\gamma}} - 1\right) \tag{53}$$

and for the actual turbine work (Eq. 52):

$$w_{turb-a} = \eta_t C_p T_3 \left(1 - \frac{1}{r^{\frac{\gamma-1}{\gamma}}}\right) \tag{54}$$

By subtracting previous equations, one obtains the actual net shaft work produced by the turbine considering the losses:

$$w_a = w_{turb-a} - W_{comp-a} = C_p T_1 \left[\eta_t \frac{T_3}{T_1}\left(1 - \frac{1}{r^{\frac{\gamma-1}{\gamma}}}\right) - \frac{1}{\eta_c}\left(r^{\frac{\gamma-1}{\gamma}} - 1\right)\right] \tag{55}$$

In a similar fashion, it is possible to show that the actual pressure ratio where the maximum actual net work takes place for a given temperature ratio, T_3/T_1, considering the isentropic machine efficiencies:

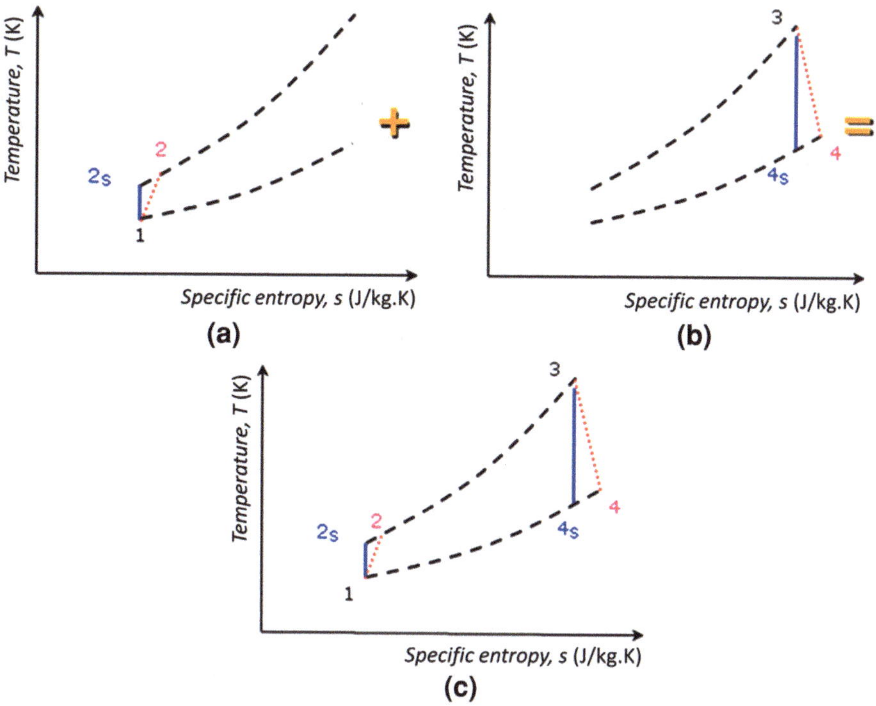

Fig. 9 **a** Actual and ideal compression work; **b** actual and ideal expansion work; **c** combination of both processes

$$r_{\max work-a} = \left(\frac{P_2}{P_1}\right) = \left(\eta_t \eta_c \frac{T_3}{T_1}\right)^{\frac{\gamma}{2(\gamma-1)}} \tag{56}$$

Similarly, one may obtain the actual thermal efficiency, $\eta_{\text{th}-a}$, as the ratio of the actual net work and the added heat.

$$\eta_{th-a} = \frac{\eta_t \frac{T_3}{T_1}\left(1 - \frac{1}{r^{(\gamma-1)/\gamma}}\right) - \frac{1}{\eta_C}\left(r^{(\gamma-1)/\gamma} - 1\right)}{\frac{T_3}{T_1} - \frac{1}{\eta_c}\left(r^{(\gamma-1)/\gamma} - 1\right) - 1} \tag{57}$$

In order to verify machine efficiencies, consider a gas turbine whose compressor efficiency is 80% and turbine efficiency is 85%. Also, consider that the minimum and maximum cycle temperatures are $T_1 = 300$ K and $T_3 = 1{,}200$ K, respectively. Figure 10a displays the ideal and actual thermal efficiencies and Fig. 10b shows the net work. As seen in those graphics, machines efficiencies are quite relevant.

Fig. 10 **a** Actual and ideal thermal efficiencies; **b** actual and ideal net work. $T_1 = 300$ K, $T_3 = 1{,}200$ K, $\eta_t = 85\%$ and $\eta_c = 80\%$

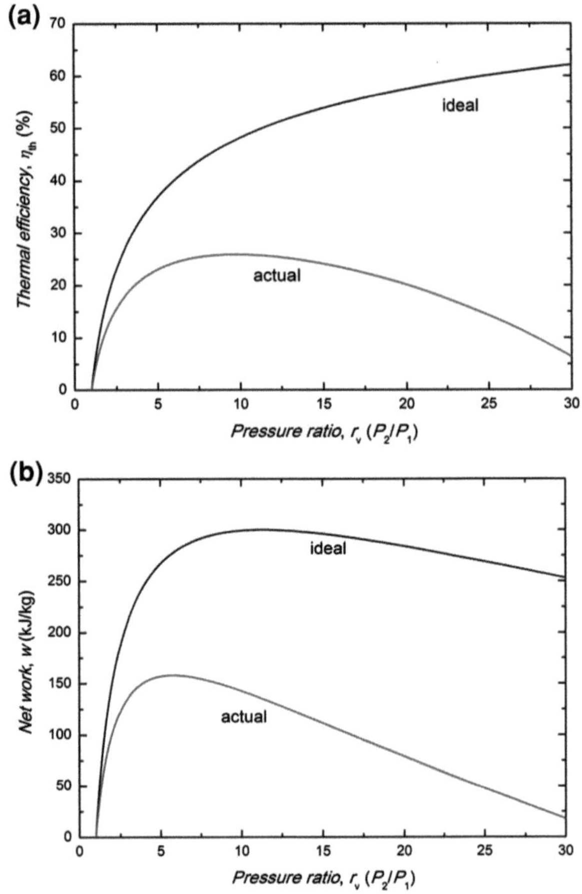

2.3 The Brayton Cycle With Regeneration

One striking point in Brayton cycle analysis is that the exhausting gas temperature is considerably high and often higher than the air leaving the compressor section. As heat will be added to the compressed air in the combustion chamber, a counter flow heat exchanger can be installed to pre-heat the compressed air by the exhausting combustion products, a process usually known as heat regeneration or heat recuperation. A schematics of such system is illustrated in Fig. 11. In the temperature-entropy diagram of Fig. 12, x represents the maximum compressed air temperature pre-heated prior to entering the combustion chamber. The area under states 2-x represents the ideal heat, and therefore, fuel saving with heat regeneration. Also, the exhausting gas will be ideally cooled to the state y in that diagram.

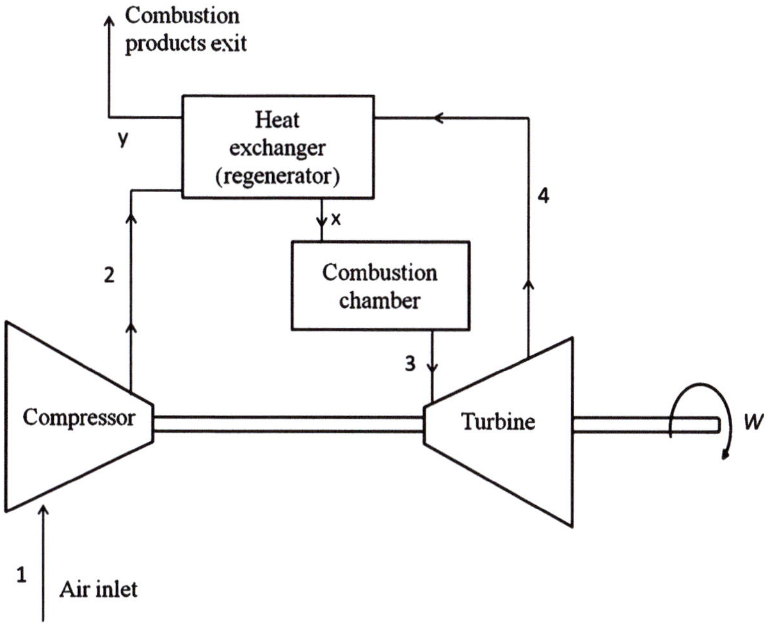

Fig. 11 Brayton cycle with heat regeneration

Fig. 12 Temperature-entropy diagram for a Brayton cycle with heat regeneration

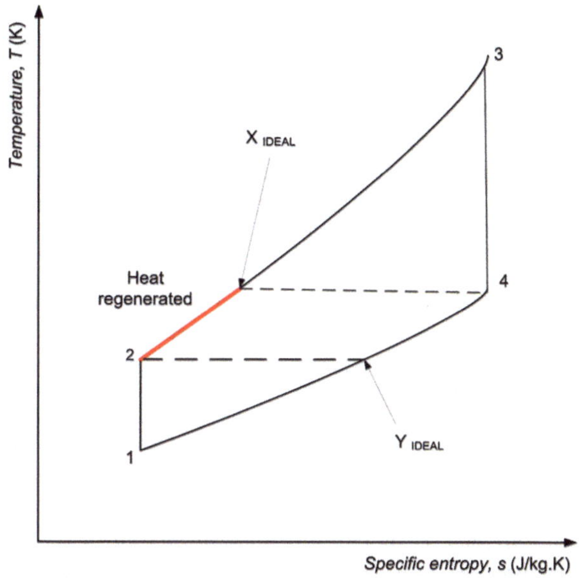

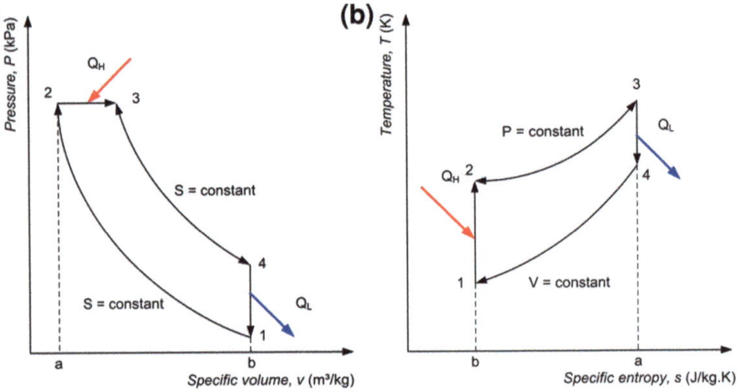

Fig. 13 Diesel cycle thermodynamic diagrams. **a** Pressure–volume diagram; **b** temperature-entropy diagram

Regeneration may be a good practice for open cycle gas turbine. In case of combined cycle configuration (Sect. 5) a previous study is required in order to find out whether there is an overall cycle improvement or not.

3 Diesel Cycle

Diesel is the air standard thermodynamic cycle used in many internal combustion engines of many small to medium thermal power plants. The working principle of an internal combustion engine is somewhat different from a closed thermodynamic cycle as it occurs also with the Brayton cycle. Working fluid composition changes from plain air to combustion products and combustion and exhaustion processes are replaced by heat transfer processes. Therefore, there is an air standard cycle that reproduces the actual machine in order to capture its main features, such as the thermal efficiency.

Figure 13 shows the two relevant diagrams for Diesel Cycle analysis. In Fig. 13a it is seen the pressure-specific volume diagram, while in Fig. 13b it can be seen the temperature-specific entropy diagram. The four ideal processes in a Diesel cycle are:

(1) process 1–2—isentropic compression, w_{comp} in the air standard cycle. Air is compressed from pressure P_1 to maximum pressure P_2. In turbocharged engines, P_1 is higher than the atmospheric pressure. In naturally aspirated engines, P_1 is the atmospheric pressure.
(2) process 2–3—heat addition, q_H, at constant pressure, $P_2 = P_3$, takes place in the air standard cycle. In actual engine, fuel is sprayed into the compressed air as its combustion takes place generating heat.

(3) process 3–4—in the air standard cycle compressed air at an initial high pressure and temperature T_3 undergoes an isentropic expansion, w_{exp}. In the actual engine, combustion products expand form high pressure P_3 to pressure P_4 generating shaft power.
(4) process 4–1—heat rejection, q_{L}, at constant volume, $V_4 = V_1$, occurs in the air standard cycle. In actual engine, the combustion products exhaust to atmosphere.

Considering the ideal processes in Fig. 13, the following energy balances can be drawn.

$$\textit{heat addition} : q_H = h_3 - h_2 = C_P(T_3 - T_2) \tag{58}$$

$$\textit{heat rejection} : q_L = u_4 - u_1 = C_P(T_4 - T_1) \tag{59}$$

$$\textit{compression work} : w_{comp} = u_2 - u_1 = C_P(T_2 - T_1) \tag{60}$$

$$\textit{expansion work} : w_{\text{exp}} = u_3 - u_4 = C_P(T_3 - T_4) \tag{61}$$

$$\textit{cycle net work} : w = w_{\text{exp}} - w_{comp} \tag{62}$$

Thermal efficiency, η_{th}, of a cycle is defined as the ratio between the cycle net work and the heat added, i.e.:

$$\eta_{th} = \frac{w}{q_H} \tag{63}$$

By substituting Eqs. 58, 59 and 62 along specific heats ratio, it yields;

$$n_{th} = 1 - \frac{q_L}{q_H} = 1 - \frac{C_V(T_4 - T_1)}{C_p(T_3 - T_2)} = 1 - \frac{T_4 - T_1}{\gamma(T_3 - T_2)} \tag{64}$$

As the process 1–2 is isentropic, then

$$T_2 = T_1\left(\frac{V_1}{V_2}\right)^{\gamma-1} = T_1 r_v^{\gamma-1} \tag{65}$$

Where, r_v is the compression ratio. In Diesel cycle fuel is injected into the combustion chamber up to a certain point known as the cutoff ratio defined by

$$r_C = \frac{V_3}{V_2} \tag{66}$$

Also, after a few manipulations it is possible to relate T_4 with T_1, which is

$$T_4 = T_1 r_C^{\gamma} \tag{67}$$

Also, considering the process 2–3 is an isobaric one, then

$$T_3 = T_2 r_C \tag{68}$$

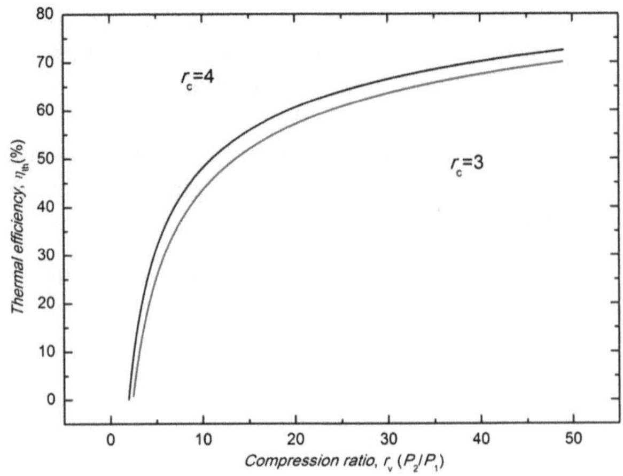

Fig. 14 Diesel cycle efficiency as a function of the compression ratio, rv, and cutoff ratio, r_c

By substituting Eqs. 65 through 68 into Eq. 64, one obtains

$$\eta_{th} = 1 - \frac{1}{r_v^{\gamma-1}}\left[\frac{r_C^{\gamma} - 1}{\gamma(r_C - 1)}\right] \tag{69}$$

Figure 14 shows the Diesel cycle efficiency as a function of the compression ratio, r_v, and cutoff ratio, r_c.

4 Rankine Cycle

Rankine cycle is the one used in steam power plants. The most common fluid used in this cycle is water, but other fluids can also be used. Lately, ROC, Rankine Organic Cycles have been devised using organic fluids, rather than water. ROC is mostly used in small to medium installations and they are usually powered by solar energy or recovered waste heat. Industrial and large thermal power plants use conventional Rankine Cycles, which are revised in this section. First, the simplest Rankine cycle is presented and the necessary variations are discussed until discussing the more commercial configurations.

4.1 The Simple Rankine Cycle

The simplest Rankine cycle is the one based on four reversible process as shown in Fig. 15a. Saturated liquid 1 undergoes an isentropic compression process to reach compressed liquid at state 2. Next, the compressed liquid is driven to the steam generator, where heat Q_H is added to obtain saturated vapor at state 4. Useful work is produced in an expansion machine, such as a steam turbine, in an isentropic process

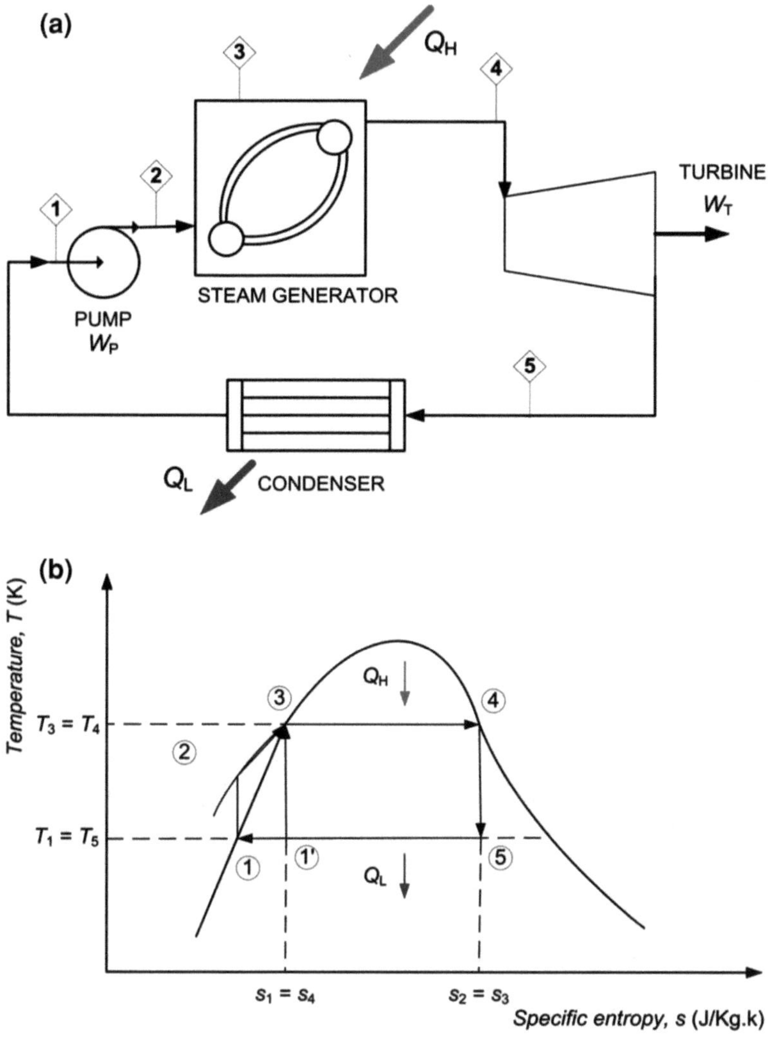

Fig. 15 **a** Four basic components of a simple Rankine cycle; **b** temperature-specific entropy diagram and Carnot cycle

yielding fluid at state 5. Finally, there occurs condensation by removing heat Q_L in the condenser to close the cycle and the fluid returns to the initial state 1. All processes are ideal. The diagram *T-s* in Fig. 15b also shows the corresponding Carnot Cycle 1′-3-4-5-1′. Clearly, one can see that the Carnot cycle has a higher thermal efficiency than the simple Rankine cycle by simply reasoning that heat is delivered to the Rankine cycle at an average temperature (between T_2 and T_H) lower than the one for the Carnot cycle (T_H).

Thermal balance around the pieces of equipment of the Rankine cycle are:

$$heat\ addition(steam\ generator): q_H = h_4 - h_2 \tag{70}$$

$$heat\ rejection\ (condenser): q_L = h_5 - h_1 \tag{71}$$

$$compression\ work\ (pump): w_p = h_2 - h_1 \tag{72}$$

$$\exp ansion\ work\ (turbine): w_t = h_4 - h_5 \tag{73}$$

$$cycle\ net\ work: w = w_t - w_p = q_H - q_L \tag{74}$$

Magnitudes in Eqs. 70 through 74 are on mass basis. For instance, if one needs the cycle total net power, W, it may be obtained according to Eq. 75, i.e.,

$$W = \dot{m} \times w \tag{75}$$

where, $\dot{m}$ is the mass flow rate. It is also a common practice to obtain the ideal pumping work by the following expression

$$w_p \cong v_1 \times (P_2 - P_1) \tag{76}$$

4.2 Rankine Cycle With Vapor Superheating

By closely examining the T-s diagram of the simple Rankine cycle (Fig. 15b), it is possible to notice that at the exit of the expansion machine (turbine) a mixture of liquid and vapor is present (state 5). Usually, a vapor quality at and below around 90% can cause damage to the turbine blades by erosion due to the impact of droplets at high velocity on them. The way to get around the blade impact problem is done by introducing a first modification on the simple Rankine Cycle. Usually, a superheater is installed at the exit of the steam generator in order to superheat the saturated vapor to higher temperatures T_6 as seen in Fig. 16a. Usually, the superheater is an additional piece of equipment integrated to the steam generator. The T-s diagram is shown in Fig. 16b.

Clearly, by heating up the working fluid to higher temperatures, a higher thermal efficiency will also be obtained without any additional increase in the working pressure. However, there is an additional cost of the superheating stage installation.

4.3 Rankine Cycle With Vapor Reheating

The previous Rankine cycle configuration can solve the problem of wet steam at the turbine exit. However, it brings about a new problem that is to superheat the turbine inlet temperature to a considerable high value. To solve this, the solution is

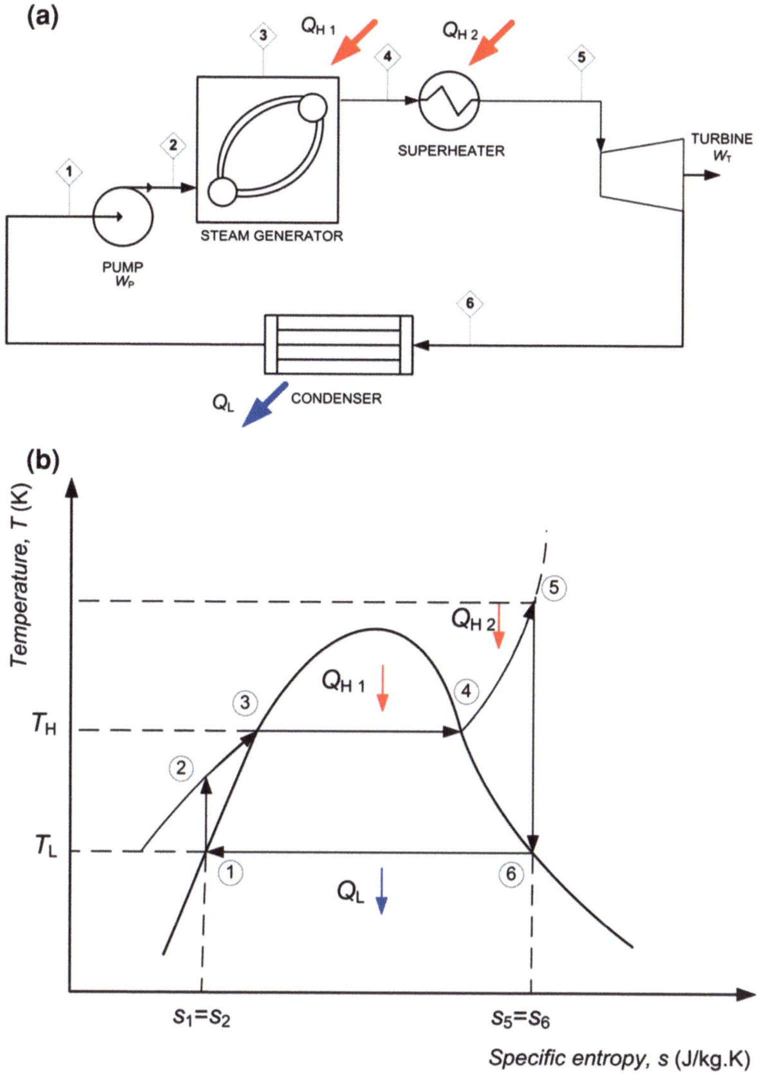

Fig. 16 **a** Rankine cycle with vapor superheating; **b** temperature-entropy diagram

to expand the vapor to a intermediate pressure and direct the vapor back to the steam generator to reheat it. Next the superheat vapor is expanded in a second stage of the steam turbine. The schematics of this configuration can be seen in Fig. 17a. The *T-s* diagram is shown in Fig. 17b. What really is done is to expand the vapor in stages so that the expansion process progresses around the vapor saturation curve in a way such vapor quality is not too high in the end of each stage. Figure 17a shows a two-stage steam turbine, but additional stages are also possible.

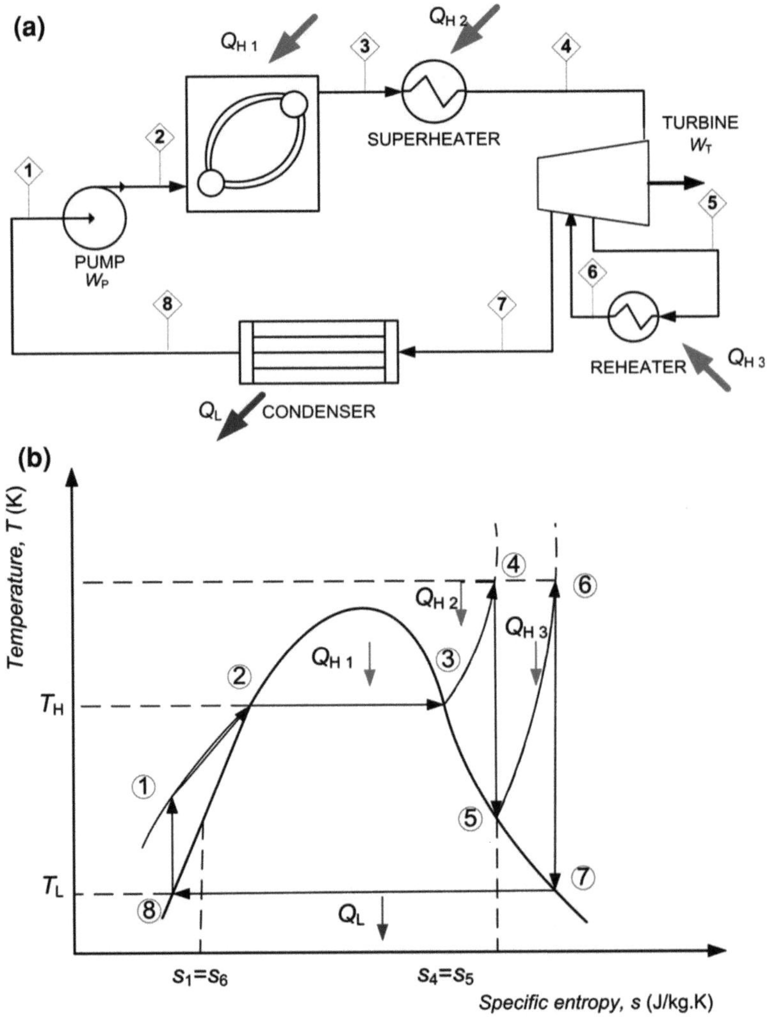

Fig. 17 **a** Rankine cycle with vapor reheating; **b** temperature-entropy diagram

4.4 Regenerative Rankine Cycle

Neither one of the previous Rankine cycle variations solves the problem of lower thermal efficiency of the Rankine cycle when compared with the equivalent Carnot cycle. Part of the problem is related to the heat addition at a low temperature liquid that enters the steam generator coming from the pump. In order to solve this, a regenerative cycle was conceived.

In a regenerative cycle, the vapor does not expand isentropically in the turbine, but as it expands it exchanges heat with the compressed liquid that travels in a

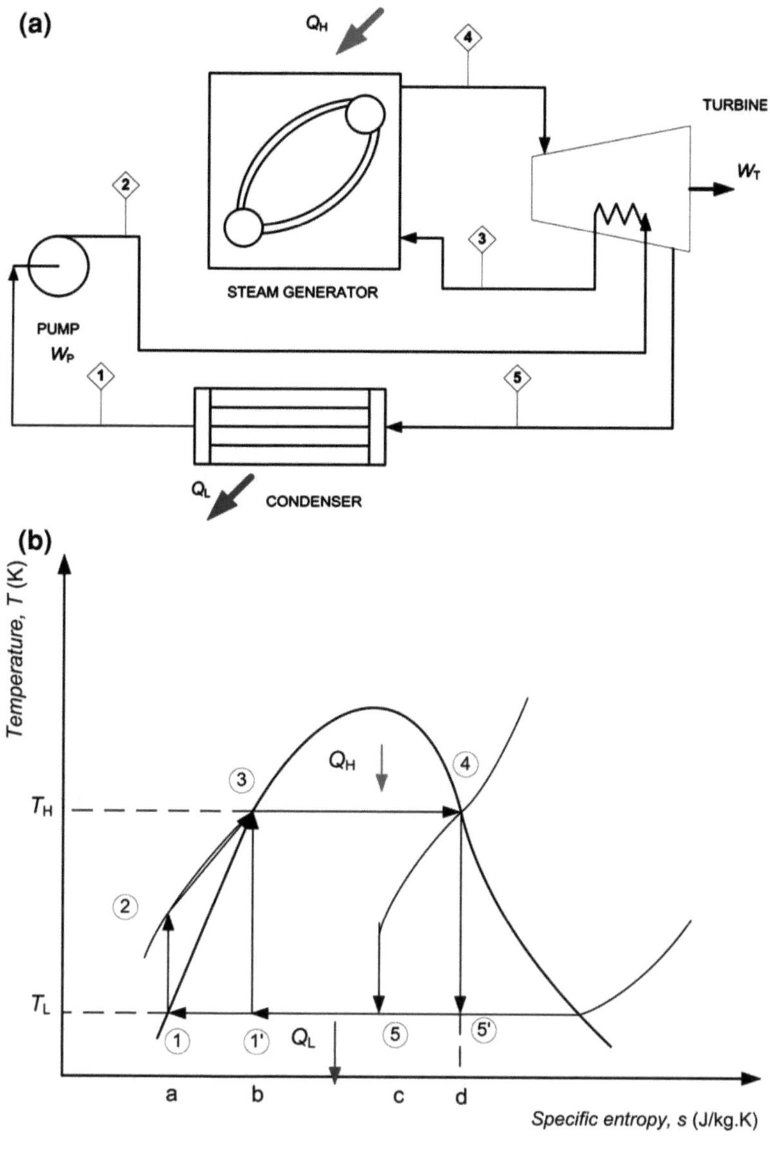

Fig. 18 **a** Regenerative Rankine cycle; **b** temperature-entropy diagram

couterflow configuration in the turbine frame. In doing so, the liquid is preheated and it will enter the steam generator as a saturated liquid. The schematic of an ideal regenegative cycle is shown in Fig. 18a. The *T-s* diagram is shown in Fig. 18b.

By examining the Fig. 18b, one can notice that ideally the regenerative cycle reproduces exactly the Carnot cycle because the exact amount of heat used to pre-heat the compressed liquid is equal to the heat lost during the expansion (areas under the curves). However, the regenerative cycle is not a practical one for at least two main reasons. First, it is not practical to design a steam turbine that it is at the same time an expanding machine and a heat exchanger. Second, the vapor quality of the vapor exiting the steam turbine is considerably lower, which can enhance the blade erosion problem. The practical solution to is extract vapor from turbine stages and mixture it with condensation water as it is examined in next section.

4.5 Regenerative Rankine Cycle With Feed Water Heating

The ideal regenerative cycle cannot be achieved in practice because of the mentioned problems. Usual practice consists of making partial expansion in the steam turbine, extract part of the vapor and mixes it with condensing water in a feedwater heater as indicated in Fig. 19a. The corresponding *T-s* diagram is shown in Fig. 19b. The thermodynamic analysis is a little more difficult because the equations now must take into account that only part of the vapor continues the expansion ($1\text{-}m_1$), while the other part (m_1) will undergo mixing with the condensing water.

A more complex cycle with three feedwater heater is shown in Fig. 20a. The corresponding *T-s* diagram is shown in Fig. 20b. An infinite number of feedwater heaters would reproduce the ideal Regenerative Rankine Cycle. This is evidently impossible In practice, six or seven heaters is usually the number of extractions and feedwater heaters used in large power plants. A detailed analysis of the optimum number of feedwater is given by Salisbury.

4.6 Losses in Rankine Cycles and Alternative Working Fluids

The most common machine to produce net shaft work in Rankine Cycles are steam turbines. As an actual device, a steam turbine is not an isentropic one and internal losses due to fluid friction with turbine blades, aerodynamics losses as the vapor flows around the blades as well as entrance and exit losses deviate the expansion process from the ideal one. In order to compute losses altogether it is usual to define the turbine isentropic efficiency as, η_t, as it was also done for the gas turbine (Eq. 52). Therefore, the actual turbine work, w_{t-a}, is obtained from the ideal turbine work, w_{t-s}, as:

$$\eta_t = \frac{w_{t-a}}{w_{t-s}} = \frac{w_{t-a}}{h_i - h_{o-s}} \Rightarrow w_{t-a} = \eta_t (h_i - h_{o-s}) \tag{77}$$

where, h_i is the specific enthalpy of the vapor at the turbine inlet an h_{O-s} is the specific enthalpy at the turbine exit in an isentropic process.

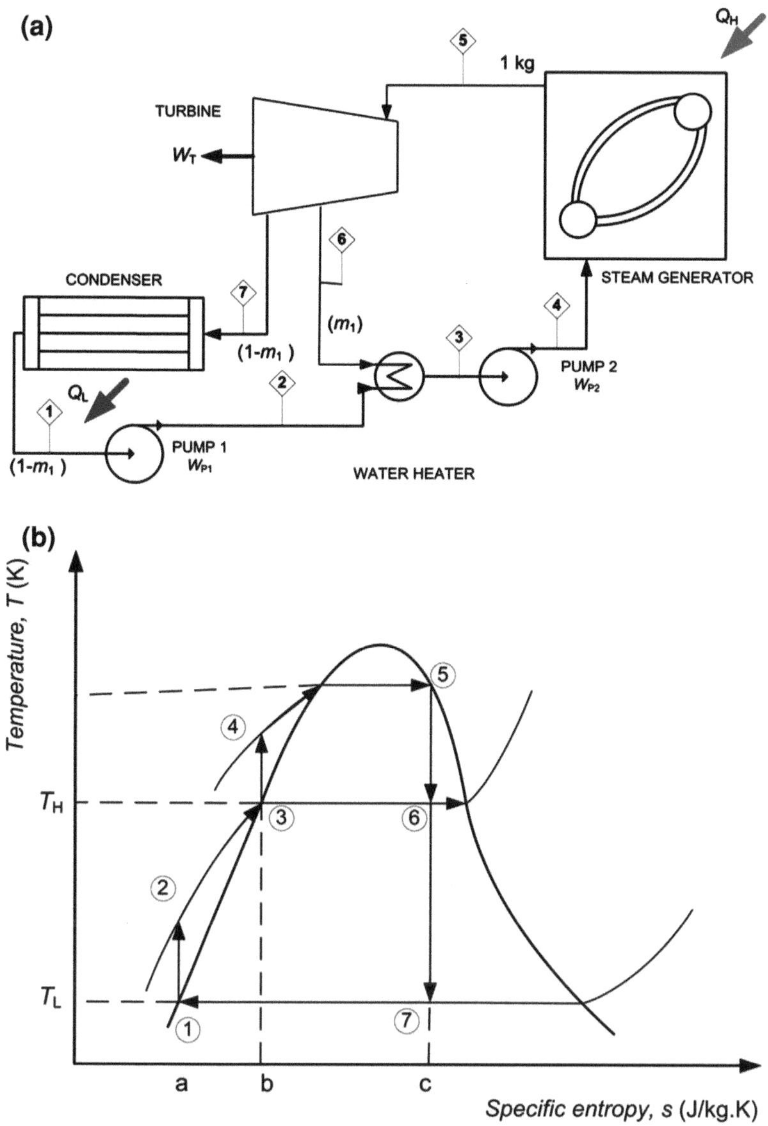

Fig. 19 **a** Regenerative Rankine cycle with feed water heater; **b** temperature-entropy diagram

Pumps are also not ideal pieces of equipment and it is required to take that into account. Other losses are associated with working fluid flow in pipes and accessories which cause distributed and local pressure losses. Finally, heat addition and rejection are not isothermal.

Huang (1988) lists a series of requirements for an ideal working fluid to operate in Rankine cycles, which are partially reproduced below.

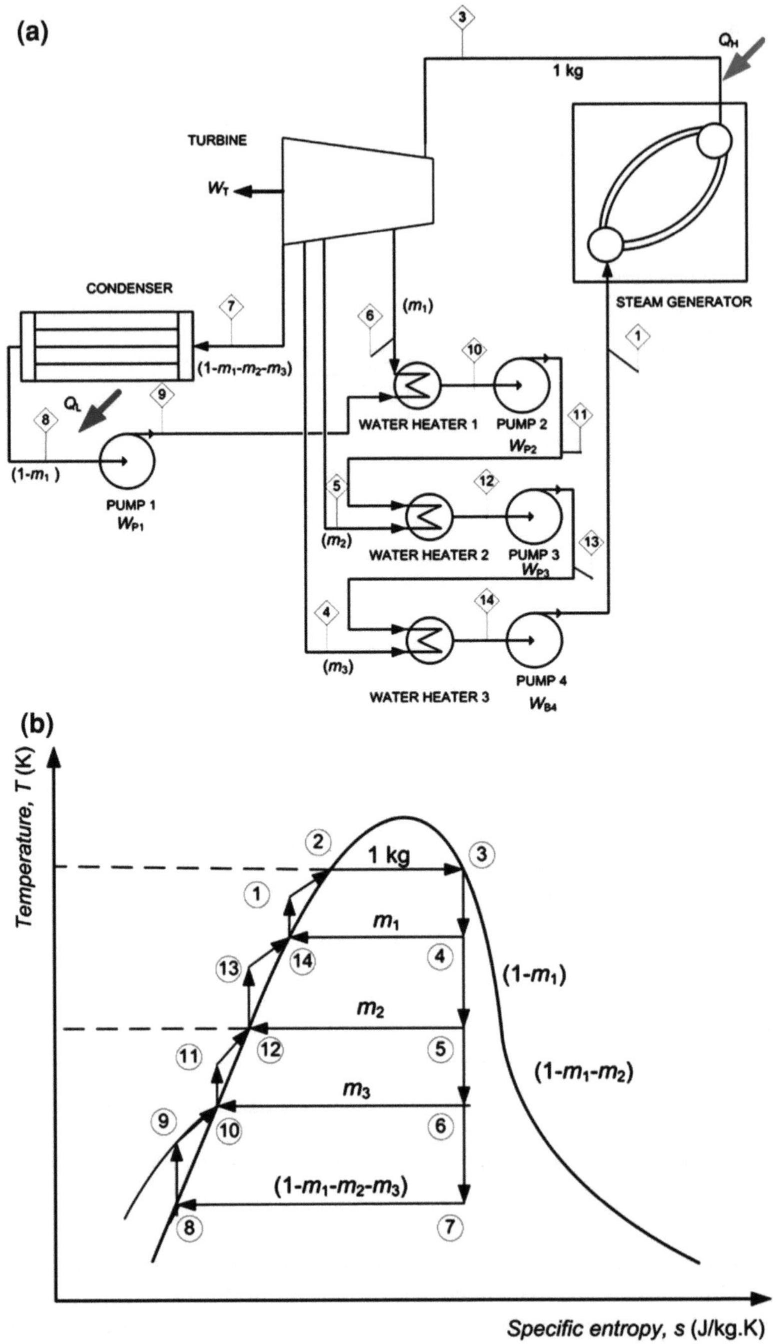

Fig. 20 **a** Regenerative Rankine cycle with three three vapor extractions and three feed water heaters; **b**temperature-entropy diagram

(1) It is desirable a high critical temperature. That would allow to work at high vaporization temperature;
(2) High vaporization enthalpy. A high vaporization enthalpy reduces the amount of mass flow rate for the same heat power added to the cycle.
(3) Condensing pressure positive. This is an important requirement, as the lower pressure in the system occurs in the condenser. External air can penetrate in the system.
(4) Positive slope of the saturation vapor curve in the temperature-entropy diagram. A fluid with a saturation curve with positive slope would avoid the use of additional pieces of equipment such as superheater and reheaters. A proper study of this class of fluids known as retrogrades can be found in Thompson and Sullivan [2];
(5) High density at operating temperatures and pressures. This would minimize the size or equipment;
(6) Nontoxic, noncorrosive and chemically stable;
(7) Low cost and widely available.

Water is the most used working fluid, but it fails to fulfill part of the previous requirements. On the positive side, water satisfies items 1 (partially), 2, 6, and 7. On the other hand, at condensing temperatures around 40°C the saturation pressure is 7.4 kPa and care must be taken to make the condenser air tight. Also, the accentuate (negative) slope of saturation vapor curve (*T-s* diagram) demands the use of superheater and reheater to avoid a high degree of wet steam at the exit of the steam turbine. Nowadays, ROC, Rankine Organic Cycles have been used in the context of solar power plants that are considering the usage of other fluids rather than water.

5 Combined Brayton-Rankine Cycle

As analyzsed in Sect. 2, a relatively high exhaust gas temperature is obtained in gas turbines. Considering economic and environmental issues it makes sense to recover such amount of thermal energy to produce other useful effect. One possibility is just to use the regeneration technique as described in Sect. 2.3. Other possibilities include; (1) using the exhaust gases to power an absorption refrigeration cycle. As a result air conditioning can be obtained for controlling the environment in working areas or, even, to cool off the gas turbine inlet air temperature, which can boost its power capacity (lower T_1 in Eq. 47); (2) using a compact heat exchanger to heat a fluid for any process purpose, including hot water and steam; and (3) using a heat recovery steam generator (HRSG) to produce steam at a temperature and pressure enough to power a steam turbine (or other expansion machine) in a Rankine cycle. Only the latter case is studied here.

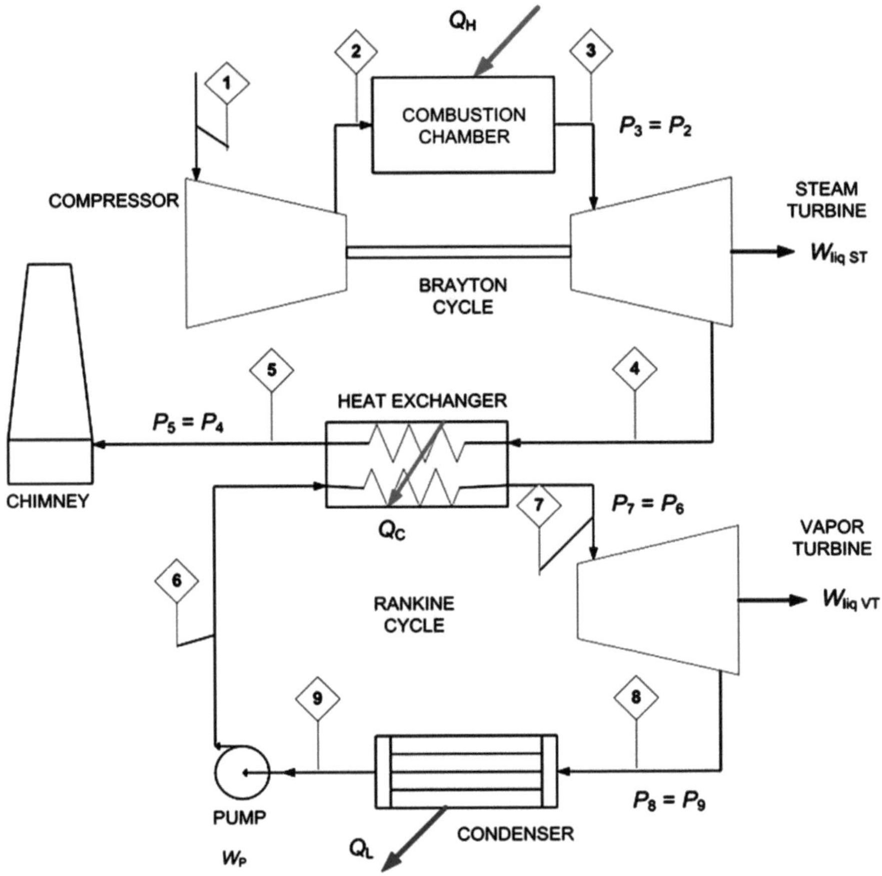

Fig. 21 Combined Brayton-Rankine cycle in 1 to 1 configuration

Figure 21 shows a schematic of a combined Brayton-Rankine cycle. Notice that steam to power the Rankine cycle is produced in the HRSG, which is driven by the exhaust gases from the Brayton cycle. As a consequence, the overall cycle efficiency is improved as seen next.

The combined cycle efficiency, η_C, is given by

$$\eta_C = \frac{W_{net}}{Q_H} = \frac{W_B + W_R}{Q_{\mathrm{H}}} \tag{78}$$

where, W_B is the net shaft power produced by the gás turbine (Brayton cycle), W_R is the net shaft power produced by the Rankine cycle, and Q_H is the thermal power due to fuel combustion. The pumping power in the Rankine cycle has been neglected. Now, by substituting the thermal efficiency of the Rankine cycle, η_R, yields:

$$\eta_C = \frac{W_B + \eta_R \times Q_C}{Q_{\mathrm{H}}} \tag{79}$$

where, Q_C is the heat load in the HRSG transferred from to exhaust gases to produce steam. Considering an ideal condition in which the exhaust gases leaving the HRSG are at a low temperature, then:

$$Q_C = Q_{\mathrm{H}} - W_B \tag{80}$$

Substituting Eq. 80 into (79) along with cycles thermal efficiencies, yields:

$$\eta_C = \frac{W_B + \eta_R \times (Q_H - W_B)}{Q_{\mathrm{H}}} = \eta_B + \eta_R - \eta_B \times \eta_R \tag{81}$$

Evidently, a combined cycle reaches higher thermal efficiencies than that of single cycle. For example. Consider a 40% thermal efficiency of a Brayton cycle and a 30% Rankine cycle thermal efficiency. The combined cycle efficiency is 58%.

The configuration in Fig. 21 is of the type 1–1, i.e., one gas turbine for one steam turbine. Other configurations are also possible depending on the individual machines. For a large gas turbine, one can think of a 1–2, i.e., one gas turbine and two steam turbines. Also it is possible 2–3, or two gas turbines or three steam turbines. It is a matter of matching machines capacities.

A combined Diesel-Rankine cycle is also possible. As a general rule, exhausting gases temperature of a Diesel engine is lower than the one from a gas turbine without a heat regenerator. Nevertheless, it is also possible to match working conditions for a lower pressure Rankine cycle.

As a final word on combined cycles, not only a simple thermal balance may result in operating cycles. The HRSG has some operational constraints depending on the temperatures and flow rates as well as the operating vapor pressure levels. It is not the goal of this chapter to discuss that problem, but approximation and pinch temperatures are issues to be analyzed on selecting power machines for operation in combined cycles.

References

1. Simões-Moreira JR (2010) An air standard cycle and a thermodynamic perspective on operational limits of Ranque-Hilsh or vortex tubes. Int J Refrig 33:765–773
2. Thompson PA, Sullivan DA (1975) On the possibility of complete condensation shock waves in retrograde fluids. J Fluid Mech 70:639–649

Analysis of Thermal Plants Configuration

Nisio de Carvalho L. Brum

Abstract Initially some thermodynamic concepts are presented aimed at a synthesis of different conceptions of thermal cycle engines. The Rankine cycle and its major components are related to the concepts previously discussed. Some operational aspects of boilers, steam turbines, condensers and cooling towers, are emphasized. The Brayton and the combined cycles are presented with emphasis on the recovery boiler.

1 Introduction

The majority of the existent thermal plants still relies on the heat produced by combustion reactions and its fuels such as, pulverized coal, natural gas or fuel oil have been used for more than one century.

Besides those fossil thermal plants are the nuclear power plants. There, the heat comes from a fission nuclear reaction. In any case the thermal plant operation may be summarized by the classical diagram presented in Fig. 1.

The temperatures of the heat source and sink are crucial parameters to establish the theoretical maximum efficiency, obtainable for any thermal engine operating between these two Thermal Energy Reservoirs (TER).

The relationship between the maximum efficiency and the temperatures is the well known equation, the Carnot thermal efficiency, based on the second law of thermodynamics,

N. de Carvalho L. Brum (✉)
LTTC—Laboratory of Transmission and Technology of Heat Mechanical Engineering Department, COPPE, Universidade Federal do (UFRJ), Rio de Janeiro, Brazil
e-mail: nisio@ufrj.br

G. F. M. de Souza (ed.), *Thermal Power Plant Performance Analysis*,
Springer Series in Reliability Engineering, DOI: 10.1007/978-1-4471-2309-5_3,

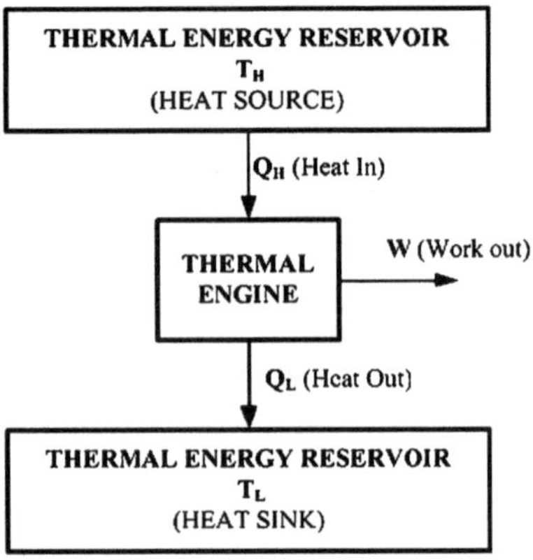

Fig. 1 Energy flow

$$\eta_{\max} = 1 - \frac{T_L}{T_H} \tag{1}$$

It is important to emphasize that the temperatures T_H and T_L are temperatures of the TERs and not of the working fluids. These fluids have varying temperatures, flowing inside the thermal engines while exchanging heat with the thermal reservoirs.

The lower temperature of the thermal reservoirs is a characteristic parameter of the atmosphere, of a river or the ocean, usual heat sinks. This temperature is an uncontrollable variable, that might be considered nearly a constant for the purpose of this introductory analysis.

A simplified expression for the heat rate, rejected by the thermal engine, $\dot{Q}_L$, is

$$\dot{Q}_L = UAf(\bar{T}_f - T_L) \tag{2}$$

where U is the averaged global heat transfer coefficient and A the required heat transfer surface.

f is a monotonous crescent function upon the temperature difference $(\bar{T}_f - T_L)$. The fluid averaged temperature $\bar{T}_f$ could be interpreted as representative of the fluid thermal process, when crossing the equipment (part of the thermal engine) where the heat is rejected to the lower temperature heat reservoir.

The Eq. 2 implies that if U is nearly a constant, smaller the temperature difference $(\bar{T}_f - T_L)$, greater the required area A, and the cost of the equipment where the heat is rejected. Alternatively greater the temperature difference more economical the heat exchanger. The increase in this difference is practically obtained increasing $\bar{T}_f$ instead of lowering T_L. And this higher $\bar{T}_f$ can be associated to a cycle with a lower maximum theoretical efficiency. In this case the temperature of the low TER will be greater than the one associated with T_L and the Eq. 1 can be used to evaluate this reduction.

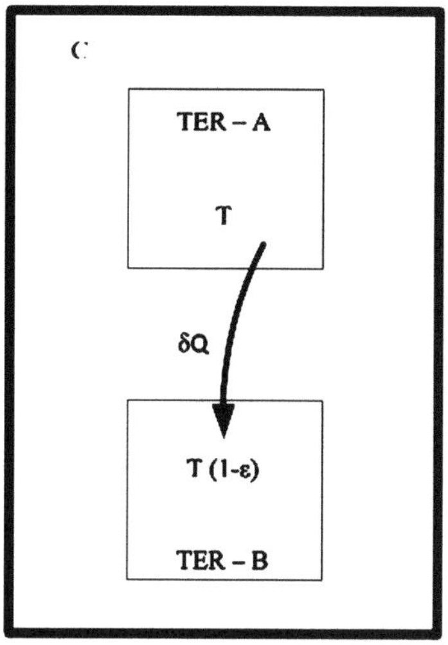

Fig. 2 Two TERs exchanging heat

A more quantitative way to show how a heat transfer with small temperature difference is more efficient is analyzing de heat transfer itself as in Reynolds [1].

Let A and B two TERs exchanging heat. C an isolated system containing only A and B as shown in Fig. 2.

The entropy variation for the system C, can be written according to Eq. 3

$$dS_C = d(S_A + S_B) \geq 0 \tag{3}$$

For the two TERs,

$$dS_A = -\frac{dQ}{T} \tag{4}$$

$$dS_B = \frac{dQ}{T(1-\varepsilon)} = \frac{dQ}{T}\left(1 + \varepsilon + \varepsilon^2 + \ldots\ldots\ldots\right) \tag{5}$$

Adding we have,

$$dS_C = \varepsilon \frac{dQ}{T} + \ldots \geq 0 \tag{6}$$

So the only way to have a reversible heat exchange is $\varepsilon \rightarrow 0$ meaning a temperature difference infinitesimal. In other words, greater the temperature difference more energy destruction, more irreversible the process and consequently less efficient.

The high temperature thermal reservoir is a model more complex to be established. The more frequent heat source, in thermal plants is the combustion reaction. It is present in pulverized coal boilers, natural gas boilers, combustion chambers in gas turbines or inside the diesel engine cylinders.

This oxidizing reaction, highly exothermic, occurs in a relative very short time if compared with the necessary time for the heat flow to be established. This results in very high temperatures for the gases that are products of this chemical reaction.

The previous discussion indicates that the temperature of the working fluid has to be as high as possible to increase the thermodynamic efficiency and to minimize the entropy increase.

After this brief introduction it will be shown how those concepts play an essential role in the thermal plants configuration.

2 External and Internal Combustion Engines

In the external combustion engine the working fluid does not participate of the combustion reaction and it is also called closed system. The most important example of this kind of engine is the Rankine cycle, the model for the steam power plant. In this case the combustion takes place in a boiler where the heat generated by the combustion is used to generate steam. There is also the heat produced in a nuclear reactor and transferred to steam generators through pressurized water.

Two variations of the Rankine Cycle are; the Kalina Cycle where the working fluid is a mixture of water and ammonia and the Organic Rankine Cycle where some special organic fluid is specified to use waste heat.

There are other examples of external combustion engines such as the Stirling and Ericsson cycles.

In the internal combustion engines, called open cycles, the air is the working fluid at least until it mixes with the fuel and undergoes the combustion. The main open cycles used in power plants are the Brayton cycle (gas turbines), Diesel and Otto cycles (reciprocating motors).

This open configuration is based on a different thermodynamic concept. In the closed system there is a unique working fluid that exchange only heat and mechanical work with the exterior. In the open configuration at its turn, there is an exchange of mechanical work, an inflow of air and fuel and an outflow of gases products of the combustion.

Although there is no heat input, we may still consider this internal combustion engine, a thermal engine. The combustion process can be seen as a transformation of internal chemical energy into internal thermal energy, in the same way a heat exchanger would do.

Regarding the heat outflow, in the Diesel and Otto engines a significant amount of heat is rejected to a cooling water, that circulates inside the engine, mainly to maintain the mechanical integrity of engine parts. The thermodynamic and mass closures necessary to maintain any engine operating steadily is done by the atmosphere, receiving the gases and renewing the air supply, and the refuel completes the "cycle", in a broad sense.

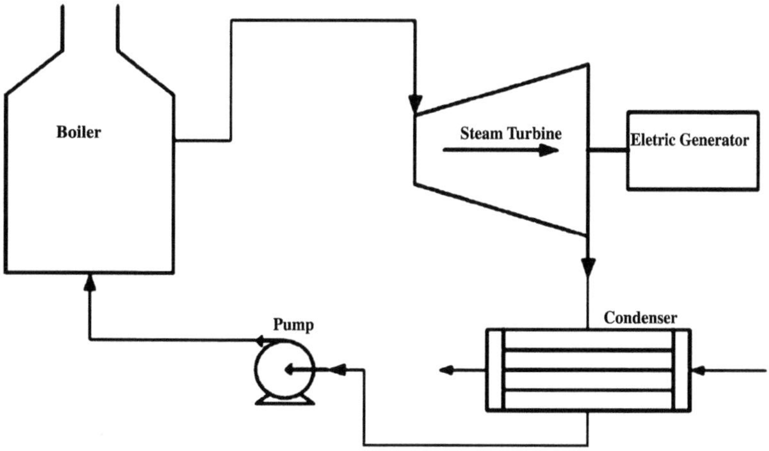

Fig. 3 The basic Rankine cycle

3 The Rankine Cycle

The basic components of the Rankine cycle are:

- Boiler or Steam Generator
- Turbine/Electric Generator
- Condenser
- Pumps

Those equipments, are interconnected as the diagram in Fig. 3 indicates.

Due the objective of this analysis, restricted to industrial power plants, we shall discuss the main components (equipments) related to this application of the Rankine cycle.

3.1 Boilers

The industrial power boilers are large and complex heat exchange equipments that are able to provide steam at high pressures and also at large flow rates.

Usually, those boilers are water tube boilers, i.e. the water flows inside several small tubes. One of the reasons is a substantial increase in the heat exchange area this will helps to control the irreversibility in the heat transfer as mentioned before.

For water the critical pressure and temperature are equal to 22,064 MPa (3,200 psia or 218 atm) and 374°C (705°F), respectively. There are boilers operating at pressures of 31 MPa and temperatures around 600°C. Those boilers are called Supercritical Boilers although the last noun might be considered inappropriate due the fact that above the critical pressure there is no boiling.

Fig. 4 Thermal circulation loop [2]

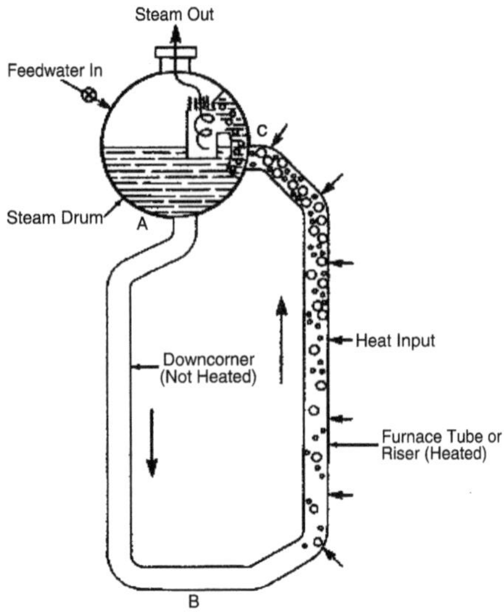

The present steam flow rate upper limit, for this kind of boiler, is around 1,300 kg/s.

One important distinction among several industrial power boiler configurations is whether exists or not a separation point between the region where the vapor is produced and the other, where the steam is just superheated.

The boilers which the separation point does not exist are called Once-Through Steam Generators (OTSG) in contrast with boilers that have a drum to separate the liquid water and steam. Figure 4 illustrates this last configuration.

The Rankine Cycle demands for high saturation temperatures for the working fluid when it crosses the steam generators. This is especially true for those which the heat source consist of the burning of fossil fuels. This need was always difficult to be fulfilled using drums.

This boiler component is a pressure vessel subjected to strong thermal and mechanical loads. Figure 5 shows this critical component.

Since the beginning of the boiler industry an alternative to this conception was sought and the OTSGs were developed in this context. In those boilers we may say, that during normal operation water molecule does not pass twice for any cross section of the water tubes or there is no water recirculation. The sketch in Fig. 6 shows a typical once-through circuit of a boiler.

The separator above is designed to remove any moisture present in steam especially in the startup and partial load operations.

These first two concepts can be summarized by saying that due the need for high temperatures/pressures in the Rankine cycle the insertion of the steam drum will become a limitation in the design of high pressure boilers. All supercritical steam generators have this once-through design.

Fig. 5 Steam drum

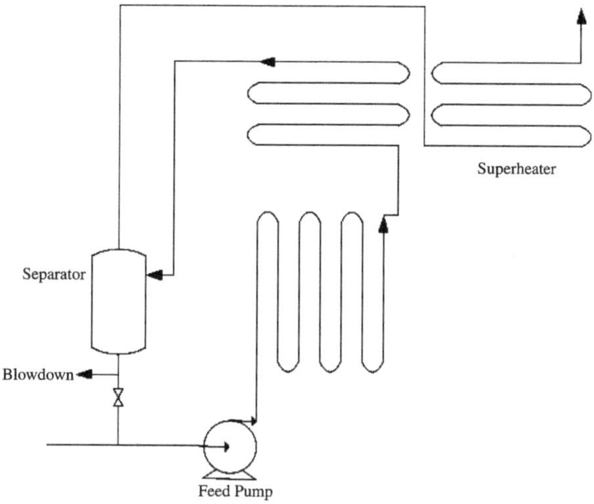

Fig. 6 Once through steam generator water/steam circuit

Another important part of the boilers is the Economizer. There, the liquid water prior entering in the evaporator (boiling region) is preheated using the hot gases after they leave the superheater.

The temperature difference of the liquid water leaving the condenser and the temperature of the flame inside the boiler is the greater possible difference occurring in the cycle. It is then necessary to reduce this difference in order to minimize the irreversibility that would occur if the liquid water enters in the boiler at such low temperature. Two actions are usually taken to achieve this goal:

- Preheat the liquid water using steam extracted from the turbines
- Using the economizer

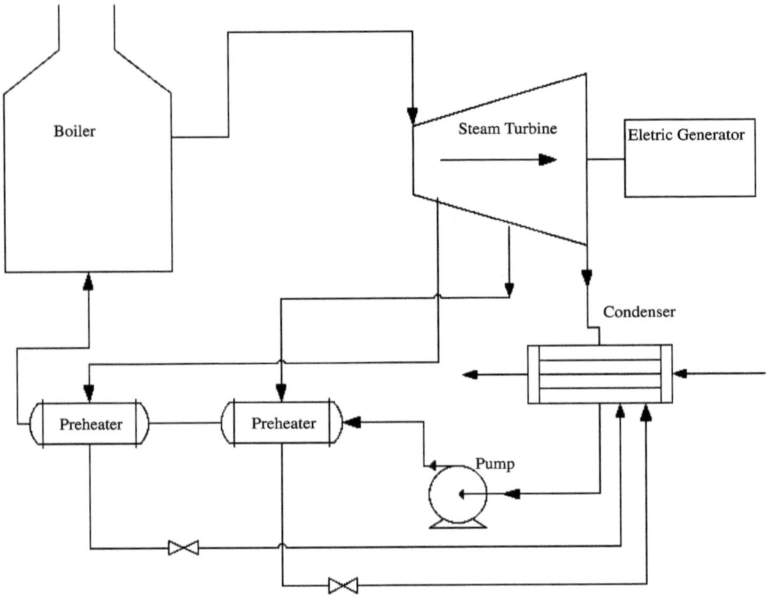

Fig. 7 Rankine cycle with preheating

The cycle in Fig. 7 illustrates the preheating with the steam extracted from the turbines. It is clear that this modification represent an increase in capital costs but the operational cost reduction pays back, even if it is used a larger number of preheaters.

The Economizer is a steam generator internal and the Fig. 8 shows one of its typical location.

Another possibility to increase the boiler efficiency is the insertion of a Reheat section. Looking to this modification from the thermodynamic second law point of view, it will control the irreversibility due to heat transfer. This will be achieved with a smaller temperature difference in the Economizer, between the liquid water and the flue gases. And the thermal energy increase in the reheat section will be internalized in the cycle to be later transformed in useful work at the turbine shaft.

The sketch bellow shows how the search for higher efficiency and less exergy destruction results in some constrains to the steam turbine design (Fig. 9).

3.2 Steam Turbines

Steam turbines are very efficient and reliable turbo machines, they convert energy in the form of enthalpy in mechanical work available as shaft torque.

Enthalpy is a function of pressure and temperature for superheated steam, in the Mollier diagram, presented in Fig. 10, is possible to notice that for a higher

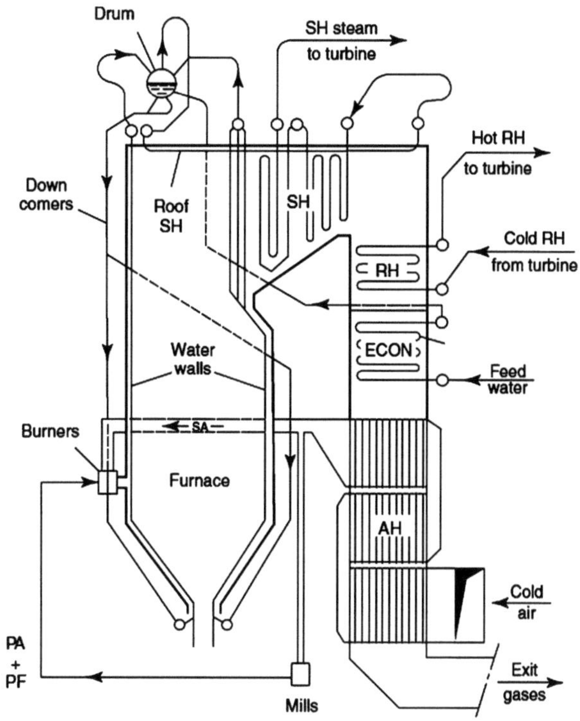

Fig. 8 Industrial boiler configuration (boilers for power and process [3]. Legend: *ECON* Economizer, *SH* Superheater, *RH* Reheater, *AH* Air heater, *PA* Primary air, *PF* Pulverized fuel

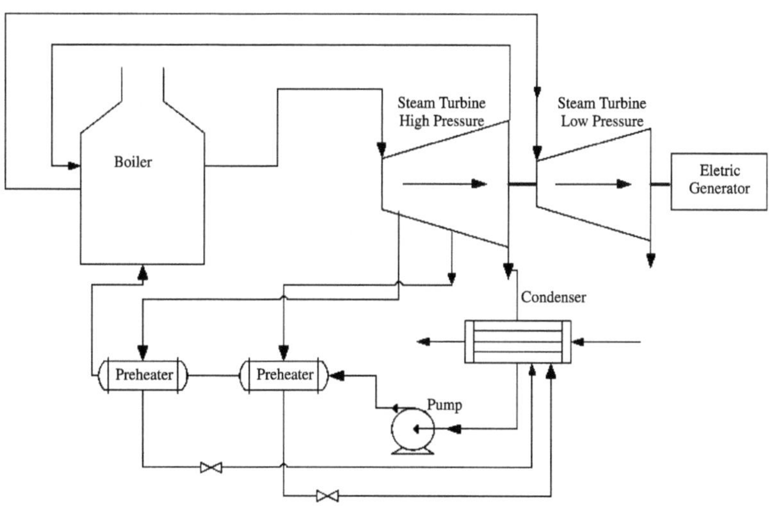

Fig. 9 Rankine cycle with feed water preheat and steam reheat

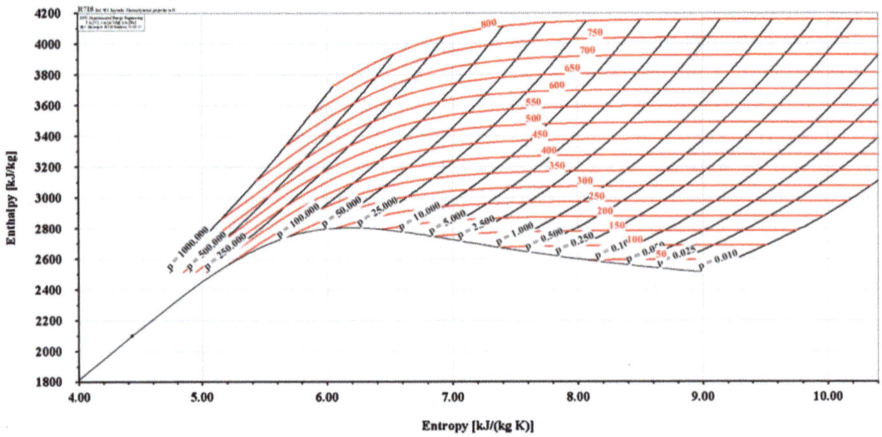

Fig. 10 Mollier chart for water

enthalpy in the turbine inlet it will be necessary high pressures (kPa) and temperatures (°C).

Inside the turbine, enthalpy is converted in kinetic energy through stationary nozzles or by the decrease in the flow cross-sectional area during the flow around the moving blades.

The high speed of the steam, after passing through the nozzles, or the high momentum obtained, will produce an impulsive force on the blade and consequently a torque on the turbine wheel. In this kind of configuration, the impulse stage, the pressure remains constant at least theoretically. Figure 11 shows a Curtis-Rateau turbine. Notice, that in all stages, when the steam flows through the moving blades, the pressure stays constant. This turbine has only impulse blading.

When this momentum variation occurs on the blades, through the area reduction the reaction of the blade's wall will also produce a torque on the wheel. In this case we will have a reaction blade, where the expansion occurs simultaneously to the torque production. Figure 12 shows a turbine, Curtis-Parson, with an impulse stage following the reaction stages.

Steam turbines that produce more than 60 MW usually have more than one cylinder. Figure 13 illustrates those turbines, called compound.

Those diagrams do not show the extraction nozzles but they are present in all configurations and they are usually located in the high pressure turbine.

3.3 Condenser

This heat exchanger operates with the exhaust steam at vacuum. And in power plants the steam flow rate are usually high, this combination leads to a large equipment which the cold fluid is water in most cases.

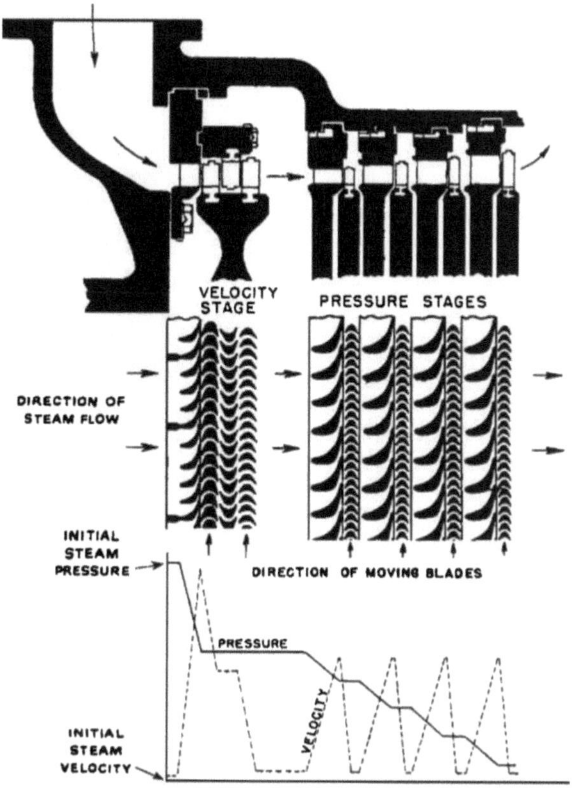

Fig. 11 The Curtis-Rateau turbine

The more frequent configuration is the surface condenser, showed in Fig. 14, where there is a metallic surface separating the condensing steam and the water. Although small the cylindrical condenser bellow shows the main parts of a condenser operating in a power plant.

One important feature is the air cooler. It is connected with the vent (air offtake) and it is just a fraction of the tube bundle, with a small tube pitch and located in a position that facilitates the non-condensable air to leave the condenser and be discarded with the help of a steam ejector.

It is included in all installations the deaerator, an equipment especially designed to remove the air from the water. Nevertheless, some air remains in a mixture with the liquid water or steam. And is inside the condenser where it separates from the steam and tends to accumulate there.

Even in small quantities, air reduces substantially the heat transfer coefficient and this can impair the condensation around the tube bundle, causing a sudden increase in the turbine back pressure.

The cooling water can be obtained from rivers, sea or recirculated water using cooling towers.

Fig. 12 A Curtis-Parson turbine

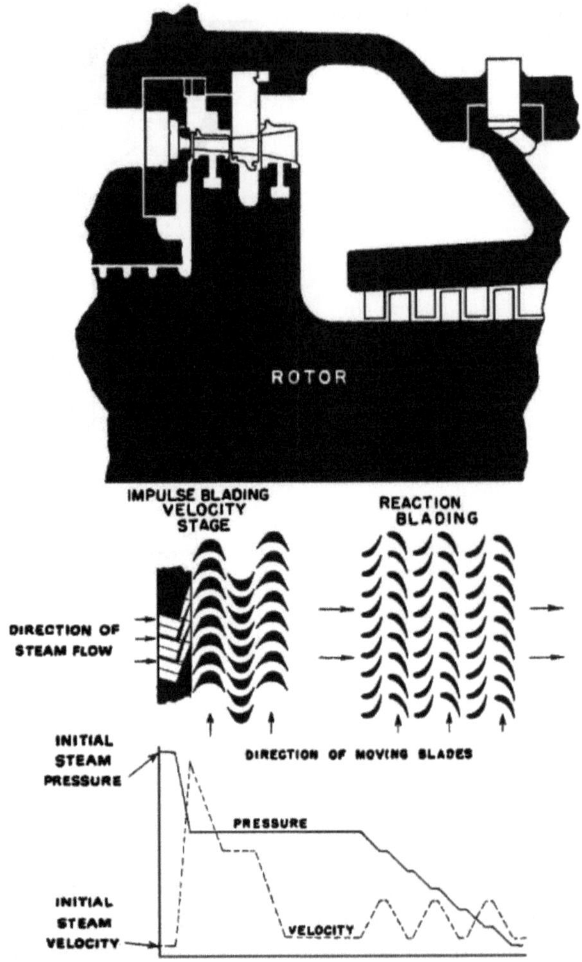

The first two options depends on the plant location and, nowadays, also costly measures to protect the river or marine environment.

The use of cooling towers reduces the water cost but it brings high capital and operational costs.

The photo presented in Fig. 15 shows a typical use of atmospheric cooling towers in a power plant.

The main heat transfer mechanism in cooling towers is the diffusion mechanism. Figure 16 shows a physical model for the energy exchange of the liquid water and the humid air.

When the water temperature is equal, in some point, to the humid air temperature, no convection heat transfer will take place. But energy is still removed from the liquid water by the diffusion of water vapor into the humid air. This last

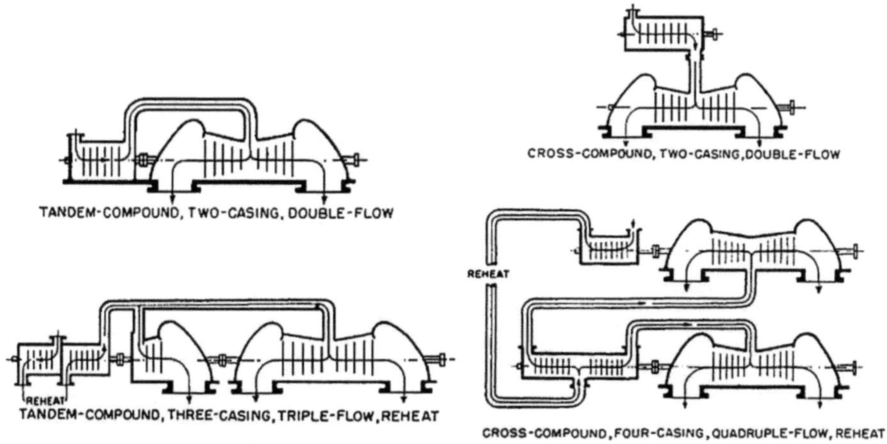

Fig. 13 Compound turbines

Fig. 14 Surface condenser

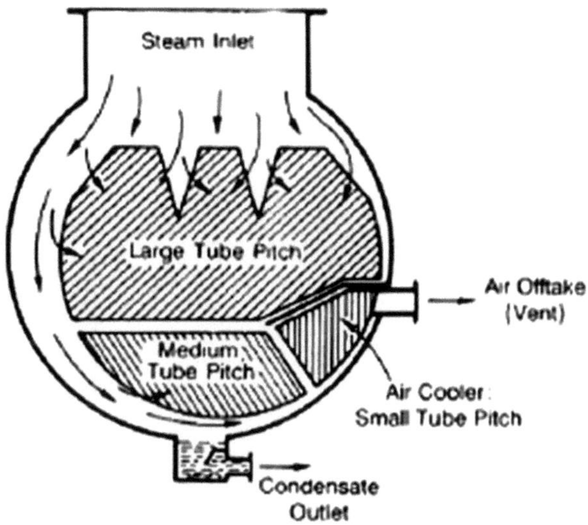

mechanism will also stop when the water vapor concentration in the humid air reaches its limit at the air temperature. The indirect measure of this concentration is the thermodynamic saturation temperature which is approximated by the wet bulb temperature. This last temperature at the tower inlet is the ultimate limit for the cooling water.

As the wet bulb temperature is usually less than dry bulb temperature the use of a cooling tower could represent an increase of the plant efficiency as the Eq. 1 indicates.

Fig. 15 Atmospheric cooling tower

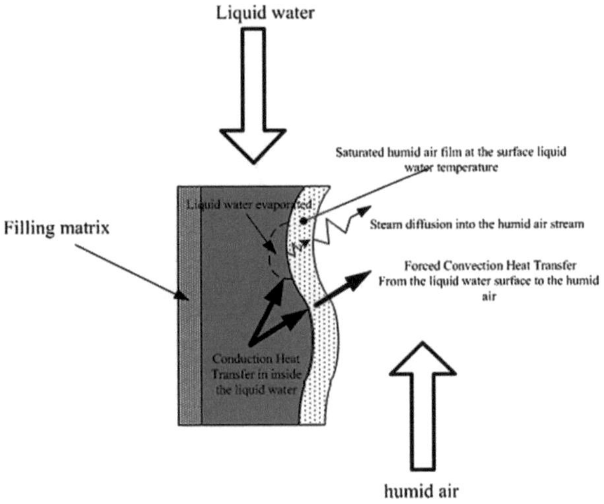

Fig. 16 Heat and mass transfer from liquid water to humid air, in a cooling tower

In some places where water is rare, there is room for air cooled condensers. Figure 17 illustrates this type of condenser,

4 The Brayton Cycle

The Brayton cycle is the thermodynamic model for the gas turbine cycle. The open configuration shown in Fig. 18 is the most common presentation of this thermal engine.

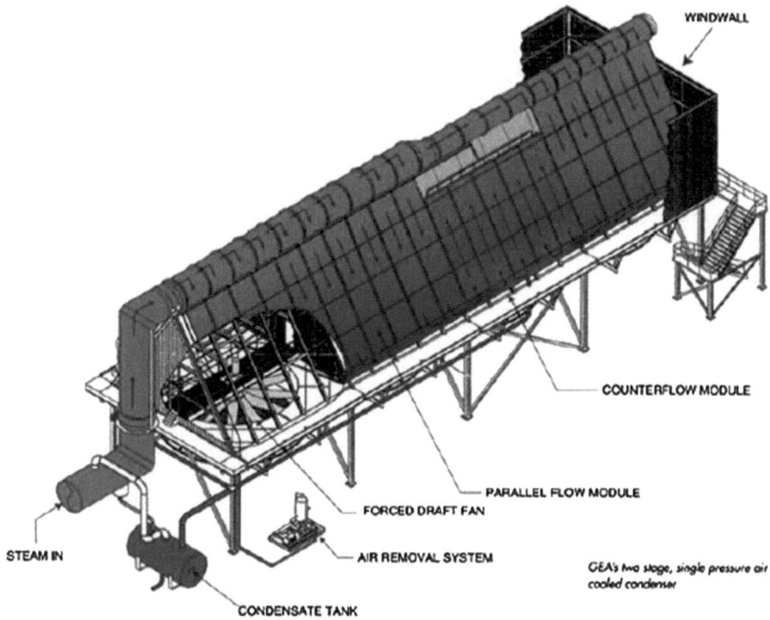

Fig. 17 Air cooled condensers (GEA heat exchangers)

Fig. 18 The open Brayton cycle

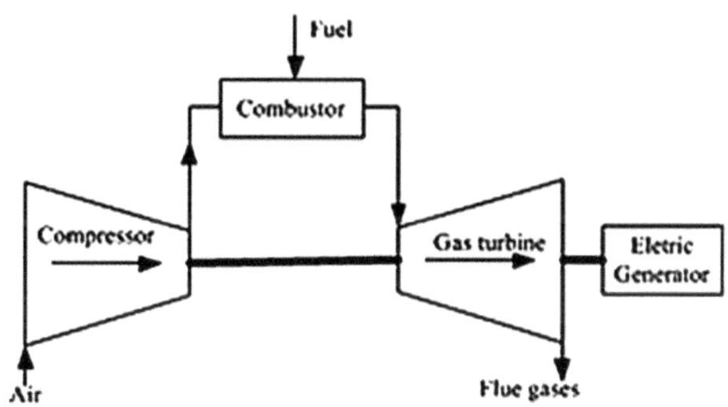

Analyzing the heat input is possible to observe a major difference with the Rankine cycle; the combustion is not external to the working fluid anymore.

Although the combustion reaction has its own irreversibility, we can imagine this cycle capable to operate with higher temperatures than the Rankine cycle. There is a potential better use of the fuel energy than we saw before. But a drawback is the use of a compressor. This component demands a high percentage of the mechanical energy produced by the gas turbine.

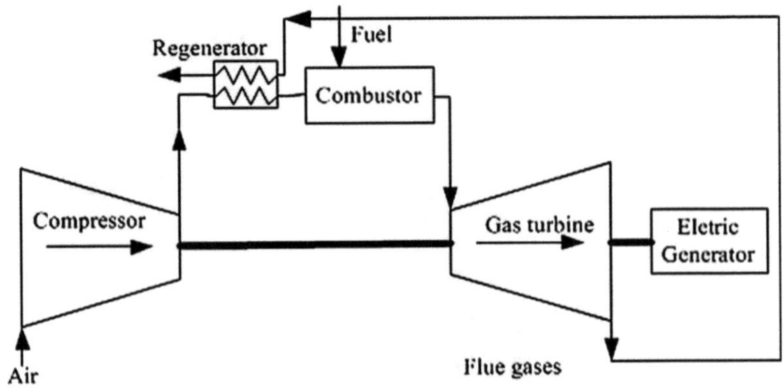

Fig. 19 Brayton cycle with a regenerator

The gas turbine is lighter and smaller than the steam cycle and this gives an enormous flexibility to the gas turbines, furthermore the gas turbine works with several fuels and has a quick starting (as low as 10 s).

There are two important categories of gas turbines:

- Industrial heavy-duty gas turbines
- Aircraft derivative gas turbines

The first of these gas turbines categories are designed to ground operation so the weight and size are not a restriction, the pressure ratio can be as high as 25:1 and the maximum turbine inlet temperature is around 1,300°C. Those cycles reach an overall thermal efficiency of 40%.

The aero derivative gas turbines are more flexible and easier to operate than the industrial ones.

The first cycle modification is the introduction of a regenerator heat exchanger, as shown in Fig. 19.

With this heat exchanger the pressure ratio could be lower reducing the compressor work. There is however, a limitation, this a gas versus gas heat exchanger, so in order to recuperate large amounts of energy, the size and the pressure drop tend to be unacceptable beyond certain power.

Another mechanism to reduce the compressor work is to cool the air while being compressed. An intercooler could do this job but this will increase the installation cost significantly.

The water injection between the low and high pressure stages will also cool the air reducing the power required by the compressor. Water is also injected in the compressor inlet and depending on the ambient temperature an increase in the net power of 8–20% is obtained combining these two injections.

There is also a possibility of a vapor injection before the combustor. This will increase the power produced in the turbine and also could control NO_x emissions.

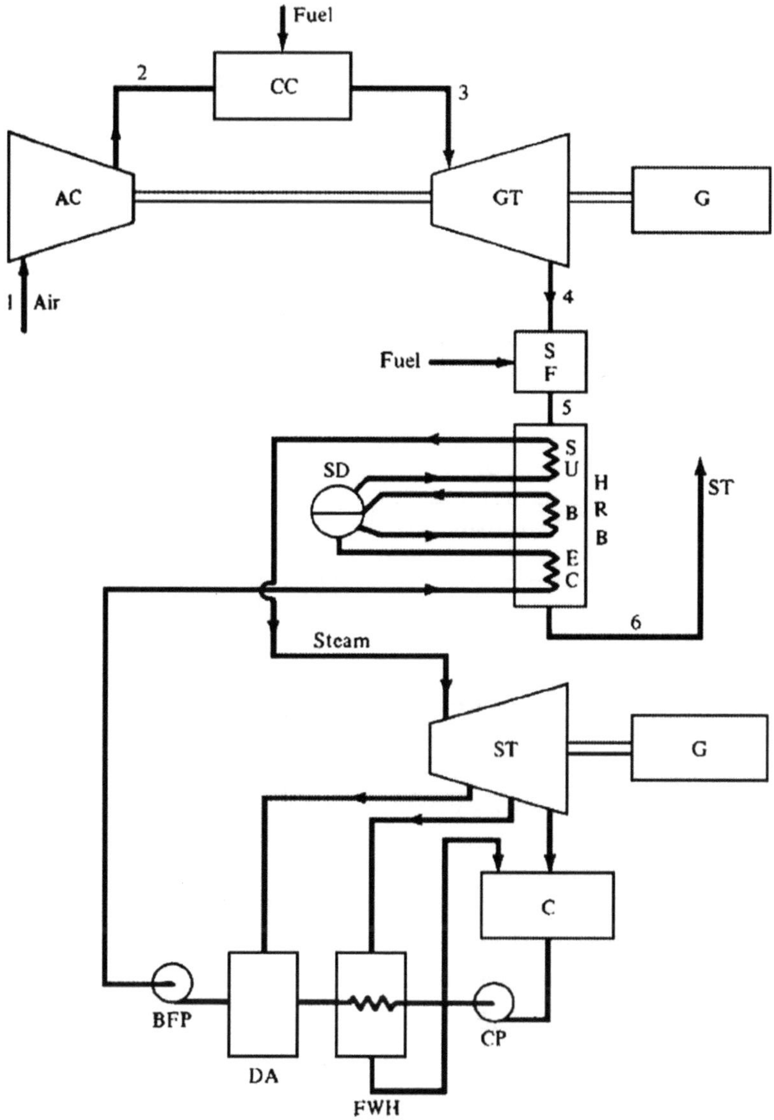

Fig. 20 The combined cycle (Power Generation Handbook Digital Engineering Library@ McGraw-Hill). Legend: *AC* Air compressor, *CC* Combustor chamber, *GT* Gas turbine, *G* Eletric generator, *SF* Supplementary fuel, *HRS* Heat recovery boiler, *SU* Superheater, *SD* Steam drum, *B* Boiler, *EC* Economizer, *ST* Stack, *ST* Steam turbine, *C* Condenser, *CP* Condenser pump, *FWH* Feed water heater, *BFP* Boiler feed pump

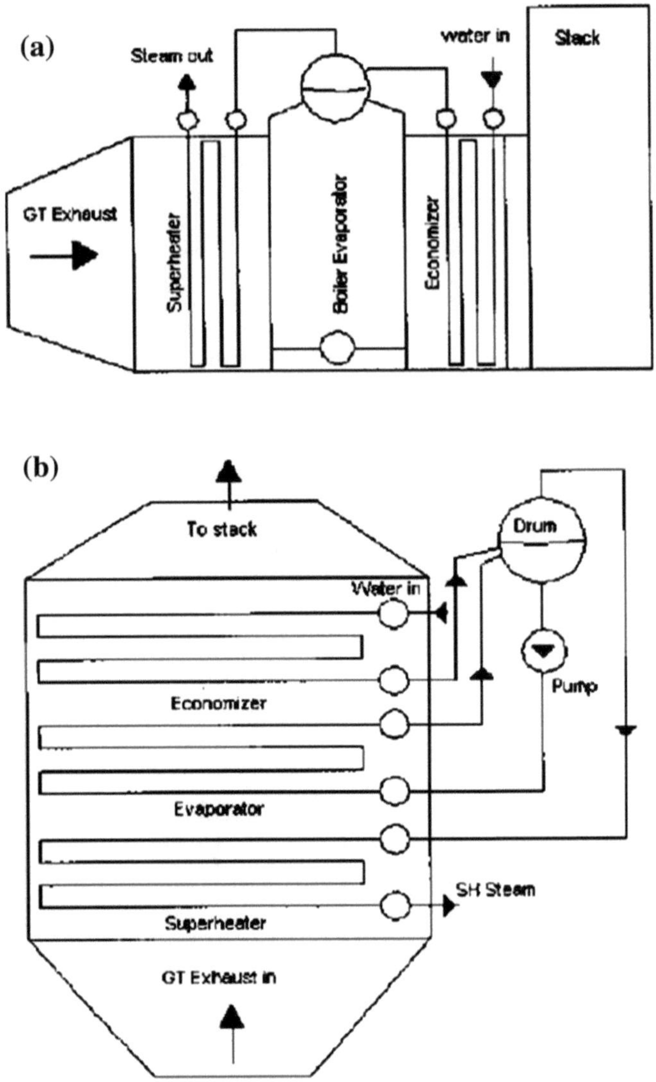

Fig. 21 HRSG configuration, (*source* Ganapathy [4])

5 The Combined Cycle

This cycle is in fact the combination of the previous cycles. It brings the ability to extract work from very temperatures (Brayton Cycle) and the capacity to produce work with a large enthalpy variation (Rankine Cycle).

The diagram presented in Fig. 20 shows a basic configuration for this cycle.

In the near future combined cycles will reach 60% of thermal efficiency, a remarkable value for this parameter.

The Heat Recovery Boiler (HRB), shown in Fig. 21, plays a crucial role in this cycle. It is a complex heat exchanger that receives the turbine exhaust gases at 540°C approximately.

Due metallurgical limitations, the gas turbine cycle uses a great amount of air in order to control the turbine inlet temperature (<1,400°C). This results in considerable amount of oxygen, ($O_2 \cong 14\%$, $H_2O \cong 6$–10%) in the exhaust gases and they can be fired with the addition of air.

Another difference in the HRBs (or HRSG Heat Recovery Heat Generator) is that the turbine exhaust gas flow, remains close to a constant, due the necessary synchronism with the electric generator. This additional firing can help to increase the steam generation in the HRB. In those boilers, the ratio of gas to steam flow varies markedly due the lower temperatures of the hot stream in contrast with traditional boilers.

With respect water circulation, those boilers have the same possibilities we have in traditional boilers, i.e. natural circulation or once-through.

References

1. Reynolds W (1968) Thermodynamics. McGraw-Hill Book Company, New York
2. Kitto JB, Stultz SC (eds) (2005) Steam, 41st edn. The Babcock & Wilcox Company, Barberton
3. Rayaprolu K (2009) Boilers for Power and Process. CRC Press, Boca Raton
4. Ganapathy V (2003) Industrial Boilers and Heat Recovery Steam Generator. Marcel Dekker Inc, New York

Fuels: Analysis of Plant Performance and Environmental Impact

Marilin Mariano dos Santos, Patricia Helena Lara dos Santos Matai and Laiete Soto Messias

Abstract This chapter discusses the impacts on operating costs, efficiency and pollutants generation of fossil fuels burned in thermoelectric power plants. The possible techniques used to reduce pollutant emission are discussed including the evaluation of their impact on the plant operation. A brief analysis of dual fuel power plant equipment is presented at the end of the chapter.

1 Introduction

Air pollution exists when chemicals are present in concentrations that are sufficient to cause harm to humans. Damage can also be derived from physical parameters such as sound and heat. The concentration of pollutants in the atmosphere depends on: climate, topography, the level of local industrial activity and population density. Pollutants are classified as primary or secondary. The primary pollutants are released directly to air. Secondary pollutants are formed in the atmosphere through reactions that occur due to the presence of proper chemical and physical conditions. Carbon monoxide, nitrogen oxides, sulfur dioxide, are primary

P. H. L. dos SantosMatai (✉)
Department of Chemical Engineering–Polytechnic School, University of São Paulo, Av Prof Luciano Gualberto, 380 travessa 3, 05508-900, São Paulo, Brazil
e-mail: pmatai@usp.br

M. M. dos Santos
Pacto Engenharia e Meio Ambiente Ltda, Rua Salto Grande, 115, 01257-020, São Paulo, Brazil

L. S. Messias
FIGENER S.A., Av. Paulista, 2644 - 1° Andar, 01310-934, São Paulo, Brazil

G. F. M. de Souza (ed.), *Thermal Power Plant Performance Analysis*,
Springer Series in Reliability Engineering, DOI: 10.1007/978-1-4471-2309-5_4,

pollutants. Low altitude ozone and sulfur trioxide are secondary pollutants. It is noteworthy that most of the air pollutants are generated through the combustion processes.

In the process of energy generation by thermoeletric power plants, fuel, air, water and chemicals are considered as being the process inputs. If not well controlled, they can trigger a series of effects on the environment that can be responsible for significant environmental impacts such as changes in local air quality. Equally to all large projects, the environmental impacts of the thermoelectric facilities are generated from their initial installation until the closure of the undertaking. However, the most significant impacts occur during operation because solid, liquid and gaseous wastes are continuously and permanently generated in significant quantities. The magnitude of the impacts depends mainly on the amount, type of waste and the ability of the environment to absorb them.

Besides the airborne pollutants, other wastes such as waste heat and noise also occur. Their nature, quantities and chemical and physical characteristics depended mainly on both, the technology and the fuel employed at the power plant. The air pollutants generated in power plants are responsible for: the increase of the concentration of the greenhouse gases and of the acidic gases that alter the pH of rain precipitation, the destruction of the ozone layer and also, the toxic gases that affect public health. The most significant liquid wastes are those generated in the systems of abatement of air emissions as wet technologies are used mainly to reduce the concentrations of acid gases. However, current practices of cleaner production (Cleaner Development Mechanism) have led to the replacement of the acid gas wet abatement systems by dry systems in order to reduce water use in industrial processes. The possible impacts arising from the generation of liquid effluents include contamination of soil, surface water and groundwater and contribute to increase the consumption of that natural resource. The solid wastes are generated in the combustion process and in the systems of dry abatement of air pollutants which use, for example, calcium carbonate to trap acid gases, or even those generated in dedusting systems. When solid fuels (coal, mainly) are employed, part of the solid waste results from fly ash that contains a portion of unburned fuel (a combustion residue) and another part, from an inorganic portion originated by the inorganics present in the fuel. If the solid wastes are not properly disposed, the resulting impacts are the contamination of soil, groundwater and surface water. Waste heat is mostly found in the exhaust gases from the combustion processes. Its most significant impact is the increase of the local temperature, thus contributing to the existence of "islands of heat". Noise is generated by the operation of equipments such as turbines and compressors. Depending on the intensity, it may contribute to increase the number of cases of occupational diseases (partial or total loss of hearing), besides causing inconvenience to people in the neighborhood. In addition to those impacts already mentioned, there may still be the visual impact due to change of the local aesthetic, caused by the installation of the power plant. It is noteworthy that all environmental aspects mentioned, cause impacts on fauna and flora in many different ways. In this context, a thermoelectric power facility that employs combustion processes can be regarded as a very important new venture. It causes significant environmental impacts that have to be properly mitigated. Among the environmental aspects, the airborne concentrations are present in greater amounts when compared to other pollutants. The magnitude of the impacts caused by the airborne concentrations of pollutants is strongly dependent on the amount, type of fuel and equipment used and also, operating conditions. Thus, for example, a plant that burns natural gas as its main fuel generates negligible amounts of sulfur oxides and particulate matter. Moreover, it can generate amounts of nitrogen oxides (NO_X) around 60% higher than plants that burn coal. Still, the use of technologies that increase efficiency of energy conversion may lead to a reduction of atmospheric emissions in relation to the energy produced. In spite of the number of possible environmental impacts from fossil fuel use for thermal generation, this chapter will address only the aspects related to the impacts on operating costs, efficiency and environment of the air pollutants.

2 Impacts of Pollutant Emission Reduction in Thermal Efficiency, Plant Operation and Capital Costs

The use of techniques of pollutant emission control in thermoelectric power plants (for both, the emission rates and the formation of pollutants) can play an important role on the overall thermal efficiency as well as on operating costs and capital expenditures. Thus, the use of techniques to prevent, minimize and/or to control the emission of pollutants is mandatory.

The pollutants that can be released from the thermoelectric power plants are: particulate matter, sulfur dioxide (SO_2), nitrogen oxides (NO_x), carbon monoxide (CO), carbon dioxide (CO_2), unburned hydrocarbons (volatile organic compounds VOCs), dioxins and furans. In addition to those pollutants, power generation is also considered an important source of CO_2 (a greenhouse gas) releasing.

The reduction of the generated emissions can be achieved through the use of end-of-pipe or post-combustion technologies that employ gas scrubbers, fabric filters and electrofilters to reduce the rates of the airborne concentrations of the produced pollutants.

Nowadays, the practices of cleaner production are privileged considering the necessity of using techniques that prevent pollution. The so-called pollution prevention techniques aim to reduce the generation of wastes that, in the case of atmospheric emissions, is the use of techniques that avoid the formation of airborne concentrations of pollutants in the process of converting the chemical energy contained in fuels.

Some of the most important devices employed to prevent pollution in thermoelectric power plants are: the use of Low-NO_x burners, the recirculation and reburning (late blight) of gases and fuel replacement by "cleaner" fuels.

For power plants, the pollution prevention measures include: fuel replacement, fuel blending, the best available technology for the fuel burned, the use of more efficient thermodynamic cycles with those that employ heat and energy generation (Cogeneration Heat and Power–CHP) and additives that reduce the emissions of NO_x, SO_2 and particulate matter.

3 Sulfur Oxides: Emission Reduction Techniques and Their Impact on Cost and Thermal Efficiency

When fuels containing sulfur compounds as contaminants are burned, substances such as SO_2, sulfur trioxide (SO_3) and sulfidric gas (H_2S) can be formed. In the combustion process, sulfur is oxidized primarily to SO_2 and only a small fraction of it (around 1–2%) is oxidized to form SO_3. The sulfur oxides last from 20 min to 7 days in the environment.

The rate of formation of SO_3 in a combustion process can be alleviated by reducing the amount of oxygen available, i.e, through the decrease of the excess of combustion air. The presence of sulfur compounds such as H_2S and SO_3 in the flue gas exhaust of power plants, besides altering the local air quality, triggering public health, altering the pH of rainfall may also be responsible for deposits in heat exchange equipment. Such deposits start the process of corrosion of metal parts and also cause obstructions in heat exchangers due to the formation of sulfates. From the operational standpoint and thermal efficiency of the plant, the presence of those deposits contributes to reduce the heat exchanging rates. Such reduction leads to a decrease in thermal efficiency and also to an increase of the consumption of the steam used for the cleaning operations that are required to restore the operating procedures. Besides the impact on thermal efficiency due to fouling, there is also the negative impact arising from the abatement systems of sulfur compounds (end-of-pipe), which consume part of the energy generated, to drive the abatement system motors. The techniques for preventing the emissions of sulfur compounds include the use of low sulfur content fuels and the reduction of the oxygen gas (O_2) available by controlling the excess of combustion air.

The choice for a technology or fuel type depends on a cost-benefit analysis of the environmental performance of different fuels, the cost of abatement equipments and the existence of a market for the sulfur byproducts that are separated by the abatement devices.

3.1 End-of-Pipe Techniques for the Abatement of Sulfur Emissions

The gas scrubbers (mostly Venturi and ring-types) are the main wet desulfurization end-of-pipe systems. Due to their high efficiency, they are widely used in large facilities such as thermoelectric power plants.

Limestone (CaCO3) is the most commonly compound used as adsorbent for desulfurization due to its availability and low cost. The byproducts are easily sold and the operating costs are minimized. Although lime (CaO) is more reactive with SO_2, it is used only in specific situations because its production process (calcination) is energy intensive.

In the desulfurization process, after passing through a heat exchanger, the flue gas is introduced into the desulfurization system where the SOx and H_2S are removed by direct contact with an aqueous solution of limestone or lime. Limestone must present a $CaCO_3$ content higher than 95%. After passing through the scrubber, the gases are sent to a demister and released to the atmosphere.

The reaction products are removed and sent for drying and other types of processing and then, sold or properly disposed.

The spray dryers are also widely used in removing sulfur gases mainly for medium-sized equipment and fuels with sulfur contents lower than 1.5%. The choice for spray dryers or for low sulfur content fuels lies on the fact that, despite high operating costs, the installation costs are smaller when compared against those of wet scrubbers or venturi-type rings. Lime, instead of limestone, is employed in spray dryers.

The low commercial value waste generated is usually a mixture of calcium sulfite, calcium sulfate and ash. Some facilities, especially those that burn coal, employ collection systems to remove particulate matter from fly ash before the gas enters the spray dryer. A byproduct with a lower content of impurities is produced.

The dry desulphurization systems involve the direct injection of dry adsorbent into the combustion chamber or duct. Powdered limestone ($CaCO_3$) and dolomite ($MgCO_3$, $CaCO_3$) are typical adsorbents. When calcium carbonate is injected into the combustion chamber, the resulting product is calcium oxide (CaO) which reacts with the SO_2 present in the flue gas to form calcium sulfite ($CaSO_3$) and calcium sulfate ($CaSO_4$); they are collected together with the fly ash in particulate matter abatement devices (usually electrostatic filters or fabric filters).

Capital costs for the dry desulfurization processes that comply injection into the combustion chamber are typically 25% higher when compared with the expenditure for the wet systems [9]. The absorption efficiency of sulfur gases for injection into the combustion chamber is about 30–50% without the recycling of the reaction products and from 70 to 80% with recycling. When injection occurs inside the duct, the abatement efficiency is about 50–80%. Concerning thermal efficiency, the systems of injection into the combustion chamber cause estimated losses of 2% in the boiler besides consuming 0.2% (on average) of the total electricity generated [9].

The technique of injecting the adsorbent into the combustion chamber can cause the deposition of solids in the heat exchange areas and also an increase in CO_2 emissions due to the calcination of carbonate. Thus, the dry adsorbent injection should be done after the heat exchangers to avoid depositions which would decrease the thermal efficiency on the power facility.

Table 1 provides information on the characteristics of the end-of-pipe desulfurization systems and the capital cost increase related to its installation.

Table 1 Performance and characteristics of end-of-pipe desulfurization systems

Desulphurization process	Characteristics	Percentage of increase on the capital cost of the plant
Wet system	Usually the flow is saturated with an aqueous solution of $CaCO_3$. The process is water intensive and produces liquid effluents that have to be treated and calcium sulfate which can be a byproduct or a residue to be adequately disposed The abatement of sulfur gases is approximately 98% and the total electric energy spent is 1–1.5%	11–14
Semi-dry system	It is widely known as dry scrubber. The abatement occurs by the absorption and reaction of the sulfur gases with CaO which is pulverized into the gaseous flow. The abatement efficiency is higher and water consumption is negligible as compared against the wet system. The total electric energy spent is 0.5–1.0%	9–12

Source adapted from World Bank Group [17]

Importantly, many of the desulfurization systems enable the regeneration of the reagent used leading to a reduction in operating costs.

The fuel replacement to reduce the emissions of sulfur gases in thermal power plants is guided by the principles of employing cleaner fuels and economic feasibility. Natural gas is preferable to oil, which is preferable to coal.

Combustion on fluidized beds is an effective form of abatement of the sulfur oxide emissions used in large power plants that burn solid fuels. Desulphurization is integrated into the combustion process. Substances such as CaO, calcium hydroxide ($Ca(OH)_2$) or $CaCO_3$ are injected into the fluidized bed to adsorb the sulfur gases formed.

4 Nitrogen Oxides: Techniques of Emission Reducing and Their Impacts on Costs and Thermal Efficiency

Due to various effects caused to the environment, over the past ten years the nitrogen oxides ($NO_x = NO_2$—nitrogen dioxide, NO—nitric oxide and N_2O—nitrous oxide) have been receiving special attention by the public and private environmental managers. Their presence in the atmosphere change the pH of rainfall (acid rain), increase the formation of low altitude ozone, cause the ozone layer depletion and contribute to increase global warming since nitrous oxide (N_2O) is a gas with a global warming potential 310 times greater than CO_2. Its residence time in the environment is 120 years. NO_2 lasts from 3 to 5 days and NO, 4 to 5 days in the environment. The mechanisms of NO_x formation, regardless

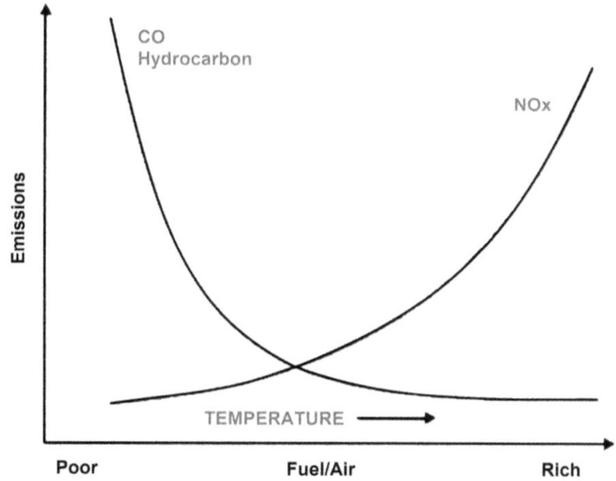

Fig. 1 The Influence of temperature on the NO_x formation rate (*Source* adapted from USEPA [16])

of the type of combustion equipment, can be summarized as: thermal nitric oxide (NO) related to the temperature of the combustion chamber, prompt NO related to the conversion of molecular nitrogen from atmospheric air with free radicals to form NO and NO related to the nitrogen present in the fuel.

The formation of thermal NO_x occurs from the dissociation of molecules of oxygen and nitrogen from the atmospheric air because the speed of the NO_x formation reactions is slower than the other combustion reactions. The NO_x formation occurs in the border regions of the flame where temperatures are around 1,700°C [18]. Equation 1 is presented to show the importance of the temperature on the formation of thermal NO. That equation quantifies the formation rate of NO from N_2 and O_2 as a function of temperature.

$$\frac{d[NO]}{dt} = 1,3.10^5.\exp\left[\frac{67.650}{T}\right].[O_2]^{\frac{1}{2}}.[N_2] \quad (1)$$

where: t: time;

NO, O_2, N_2: concentration, mols/cm^3; and,

T: temperature (K).

The concentration of the NO_x formed is strongly dependent on temperature and to a lesser extent, on the concentrations of N_2 and O_2. It indicates that a way of controlling the NO_x formation is to avoid temperature peaks. Figure 1 shows a diagram that illustrates the influence of temperature on the formation of NO_x in gas turbines.

According to Fenimore [7], the formation of prompt NO occurs through the reaction of molecular nitrogen with radicals such as CN (cyanide) and substances as HCN (cyanidric acid), and is limited to the area of the flame front. The mechanism is strongly dependent on the existence of hydrocarbons.

The prompt mechanism, which predicts the reaction of molecular nitrogen with free radicals was proposed by Fenimore who could not employ the model proposed by Zeldovich to predict the rates of NO_x formation in the regions neighboring the flame. Fenimore considered that the reaction rates would be extremely fast and also, that considerable amounts of free radicals of hydrocarbons should be present. Thus, he proposed a mechanism involving the radicals present in the flame front of the combustion of hydrocarbons with nitrogen species. Subsequent experiments performed by Blauwers et al. [2] indicated that the CH and CH_2 radicals play the most important role for that mechanism.

Carvalho and Júnior [4], Bowman [3] and Miller and Bowman [11] reviewed scientific papers published on the subject and showed that there are studies reporting that, part of the NO formation in the flame region, does not follow the mechanism described by Fenimore. According to those authors, some studies show that there is an equilibrium between the concentrations of elementary oxygen (O) and hydroxyl (OH), which accelerates the reaction rates from reaction 1 to 3, especially in poor combustion conditions. According to Carvalho, there is no consensus among researchers upon the NO formation mechanism represented in the following reactions:

$$N_2 + O \rightarrow NO + O \quad \text{Reaction} \quad 1 \tag{2a}$$

$$N + O_2 \rightarrow NO + O \quad \text{Reaction} \quad 2 \tag{2b}$$

$$N + HO \rightarrow NO + O \quad \text{Reaction} \quad 3 \tag{2c}$$

Carvalho and Júnior [4, p. 70] states that prompt NO only overlaps thermal NO in rich combustion conditions. In poor combustion, thermal NO corresponds to 70% of the NO formed. Thus, a way to control the NO formation is to work under impoverished combustion by staging the combustion of the fuel. The formation of NO_x from the nitrogen present in the fuel composition occurs from the thermal decomposition of nitrogenated organic compounds into substances such as ammonia (NH_3) and HCN whose oxidation rates are comparable in magnitude to the combustion oxidation reactions. Unlike thermal NO_x, the formation of nitrogen from the fuel does not show a strong dependence on temperature. The main factor governing its formation is the availability of oxygen in the flame region and also, the availability of nitrogen compounds.

A control procedure is to use of fuels with low or no nitrogen in their composition. Most of the fossil fuels present variable amounts of nitrogen. Fuel oil produced in oil refineries, coal and natural gas present contents of respectively, around 2%, from 1 to 2%, from 0.5 to 1% of nitrogen compounds.

Figure 2 illustrates the importance of fuel in the rate of NO_x formation as a function of the nitrogen concentration in the fuel and the temperatures reached during the combustion process.

The analysis of the NO_x formation mechanisms show that the main factors that contribute to NO_x formation are temperature, air excess, residence time of the gases at high temperatures, the mixing rates of fuel and air and the nitrogen content in fuel. Among these factors, temperature is the most critical, so the strategies that somehow reduce the temperature of the combustion flame can positively impact

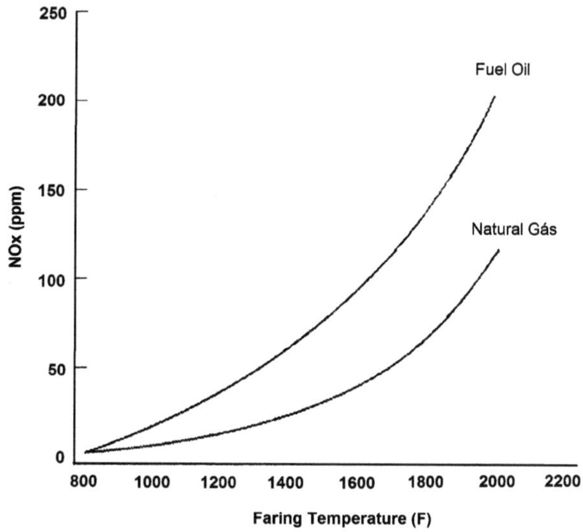

Fig. 2 Relationship between fuel with gas temperature and No_x formation rates (*Source* USEPA [16])

the reduction of NO_x formation. Thus, the NO_x emissions are likely to be controlled not only by techniques such as end-of-pipe, but also by pollution prevention techniques that reduce the formation rate through changes in process or combustion equipment.

In spite of the formation mechanism of NO_x being alike for boilers, furnaces and turbines, the techniques of controlling NO_x emissions used in thermoelectric power facilities may impact in different ways on each of these devices, either in thermal efficiency or in operating costs and capital.

4.1 Pollution Prevention Techniques to Reduce the Emission Rates of NO_x in Boilers

The mechanisms of NO_x formation indicate that the production control must include strategies that minimize the combustion temperature and the residence time of the species in the peak temperature regions and also, reduce the available oxygen without affecting combustion quality.

The pollution prevention techniques to the NO_x emissions involve from no cost actions as an adjustment of the excess of air to costly devices such as the installation of costly Low-NO_x burners. Their design provides the reduction in pollutant rates. Figure 3 shows the main strategies for the prevention of NO_x emissions in power plants. The control of NO_x formation by reducing the air excess is a relatively simple procedure to be operationalized. It allows the reduction of the amount of available oxygen in the flame region where temperatures are high and also also reduces the peak temperature and the availability of nitrogen in the flame. Those are

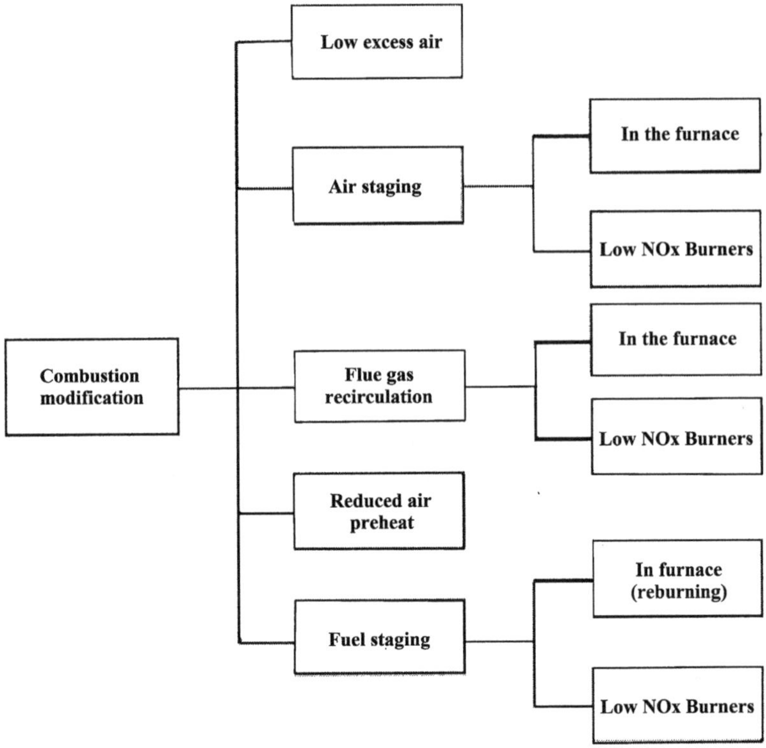

Fig. 3 Strategies for the prevention of NO_x emissions in power plants (*Source* adapted from Rentz et al. [13])

the determining factors in the formation of thermal NO_x. The reduction of the air excess, besides reducing the NO_x formation, can also play a positive impact on the thermal efficiency of boilers if the stoichiometric ratio is maintained at levels that provide complete combustion. The control of the air excess within appropriate levels that allow complete combustion is essential, because a low air excess can lead to an increase of the carbon monoxide (CO) and unburned carbon levels, resulting in poor efficiency. The costs of implementation of such measure is insignificant, since it only requires an adjustment of the combustion parameters.

Staged combustion provides rich and poor regions inside the combustion chamber thus preventing the achievement of favorable conditions of temperature and concentrations that allow the production of NO. Staging is the fractionated injection of fuel or oxidant (combustion air). Usually, the primary zone is a fuel-rich area into which around 70% of the stoichiometric air is injected, that is, there is low oxygen availability and therefore, a reduced interaction between the molecules of O_2 and N_2. It is a low temperature region.

The effectiveness of NO_x reduction in the rich combustion zone is restricted by the same factors that limit the use of the technique of low air excess, i.e, the formation of CO and partially oxidized organic compounds. As these compounds

Fig. 4 Denitrification rate as a function of the late blight rate in boilers (*Source* adapted from Rentz et al. [13])

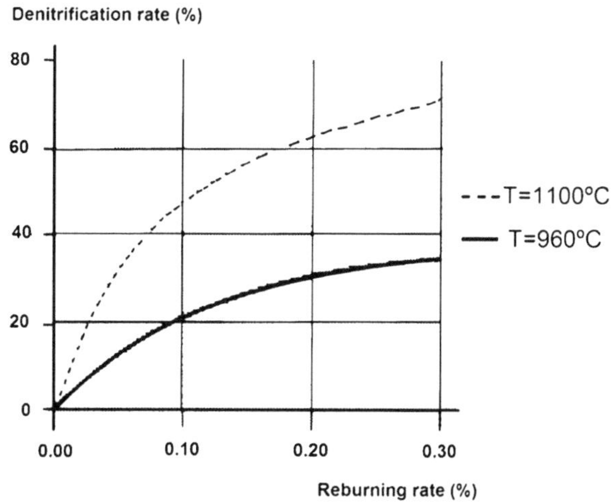

are formed, they may be deposited and cause obstructions in the boiler tubes, causing loss of thermal efficiency.

The control of NO_x formation through the recirculation of the combustion gases into the flame region is an action that reduces temperature and the concentration of oxygen available in the core of the flame thus contributing to the reduction of NO_x formation. Amounts of 10–30% of the exhaust gases are injected along with the secondary air in the combustion chamber.

NO_x reductions of 40–50% can be achieved by the recirculation of 20–30% of the exhaust gases from a gas or oil boiler. However, the increase in recycling rates to levels close to 30%, may generate combustion instabilities leading to the formation of CO and soot.

The recirculation of gases implies costs that are higher than those spent for the changes of excess air or staging. The difference relates to the need of installing ducts and fans that operate at elevated temperatures.

The technique of reburning (late blight) or fuel staging consists of injecting fuel and air into different regions of the combustion chamber. The goal is to convert part of the NO formed in the main flame region to N_2.

The reburning technique involves the combustion in three different zones: the primary combustion zone, where 80–85% of the fuel is burned in an oxidizing or slightly reducing atmosphere; the secondary combustion zone, also known as late blight zone, where additional fuel is injected to generate a reducing atmosphere. In this region the hydrocarbon radicals produced react with the NO_x formed in the primary zone. Other compounds such as ammonia are also generated in this area. Several reburning fuels can serve as fuel blight (pulverized coal, fuel oil, natural gas, etc.) for the reburning technique. However, natural gas has been widely used due to its intrinsic property of not having nitrogen in its composition. Figure 4 shows a chart that relates the rate of denitrification as a function of the rate of late blight in two different temperatures.

Reburning can be implemented for all types of fossil fuels and also, in combination with other NO_x reduction techniques. However, the technique requires large volume combustion chambers, which can be restrictive in retrofit situations when there is limited space in existing facilities. The main issue concerns incomplete combustion which requires large volume combustion chambers that allow longer residence times leading to complete combustion.

The costs of reburning depends on the structure of the boiler and the fuel used. The use of an auxiliary fuel, such as natural gas, increases costs, but it is considered as being a part of the fuel. Estimates indicate that the implantation of a 250 MW boiler with reburning is on the order of \$ 14/kW thermal. Estimates also indicate that the operating cost of a boiler with reburning is about twice as much the operational cost of a boiler provided with a Low-NO_x burner.

The principle of the NO_x reduction in Low-NO_x burners lies in the fact that it provides flames in two distinct zones. In the primary zone, temperatures are high and the atmosphere is reductive, located at the root of the flame. In the secondary zone, located at the end of the flame, the temperatures are low and the atmosphere is oxidizing. Most of the NO is generated in the primary zone and its concentration increases exponentially with temperature. However, the contribution of the secondary zone is greatly reduced.

The creation of the two zones in the low-NO_x burners are achieved mainly by modifying the means of introducing air and fuel in order to delay the mixture, thus reducing the availability of oxygen and the peak temperature of the flame.

The Low-NO_x burners slow the conversion of the nitrogen present in the fuel to NO_x and also, the production of thermal NO_x without decreasing the combustion efficiency. Currently there are several burners of that type, which present different control strategies incorporated in their design: the staging of air, flue gas recirculation and fuel staging. The use of Low-NO_x burners incurs additional operating costs due to, for example, the increase of pressure loss in air ducts. In many cases, improvements in the coal pulverization system are required for the boilers that burn that fuel.

The analysis of the techniques for reducing of the NO_x formation rates shows a remarkable disadvantage which is the possibility of increasing emissions of other pollutants such as CO and hydrocarbons (HC), and that is due to either, low temperatures or rich combustion. For those reasons, if the prevention techniques are not well controlled, a significant loss of thermal efficiency can take place.

If the pollution prevention measures are not sufficient for the rebate or even to cause significant loss of efficiency, the simultaneous employment of other measures such as end-of-pipe for the control of NO_x emissions should be considered. Table 2 provides information on key techniques, their advantages and disadvantages for reducing the NO_x formation rates, used mainly in boilers.

The percentage reductions shown in Table 2 are estimates, considering that each power plant has its own characteristics.

Table 2 Techniques for the reduction of NO_x formation: advantages and disadvantages

Technique	Advantages	Disadvantages	Percentage of reduction	Applicability
Excess of air reduction	Thermal efficiency is improved. No need of investments	Low efficiency on NO_x reduction	Gas 16–20 Oil 16–20 Coal 20	To all types of fuels
Staged combustion	Low cost and compatibility with other techniques	Average cost installation	Gas 30–40 Oil 30–40 Coal 30–50	To all types of fuels
Gas recirculation (30%)	Significant reductions	Flame instability and high cost	Gas 40–50 Oil 40–50 Coal NA	To fuels with low nitrogen content
Reduction of the pre-heated air flow	Potential for significant reductions	Reduction on thermal efficiency	Gas 15–25 Oil 15–25 Coal 15–25	To fuels with low nitrogen content

Source adapted from USEPA [16]

4.2 End-of-Pipe Techniques to Reduce the Emission Rates of No_x in Boilers

The end-of-pipe technologies applicable in the case of the NO_x, involve the injection of ammonia, urea and other compounds that react with the NO_x to reduce them to molecular nitrogen. The most important are: Selective Catalytic Reduction—SCR and Selective Non-Catalytic Reduction—SNCR.

The Selective Catalytic Reduction (SCR) systems are widely used in large installations such as power plants. The process is based on the selective reduction of the nitrogen oxides with ammonia or urea in the presence of a catalyst.

Despite showing an abatement efficiency of 95%, the process has the potential disadvantage of releasing ammonia due to the incomplete reaction of ammonia with nitrogen oxides either by the loss of conversion efficiency of the catalyst, or by the excess of ammonia. The incomplete reaction of ammonia with NO_x is known as ammonia slip (NH_3 slip). Besides the possibility of releasing ammonia, the NH_3 slip process may also cause fouling on the air preheaters or on the catalyst surface due to formation of ammonium sulfate an and also, it promotes the presence of ammonia in the in wastewater from the systems and in the cleaning effluents from the heat exchangers.

The investment costs in boilers vary depending on the type, volume of gas to be treated and the conversion rate of NO_x expected. The operating costs are strongly dependent on the lifetime of the catalyst as well as on the consumption of the reducing agent and energy consumption of gas to reheat the engines.

The Selective Non-Catalytic Reduction Systems (SNCR) were developed to complement the techniques for the reduction of NO_x formation rates, for example, the Low-NO_x burners.

Table 3 Performance and characteristics of the End-of-Pipe denitrification systems

Process type	Characteristics	% of increase on capital cost of the plant
SCR	-Reduction rates of 80–95%	
	-Electric energy consumption around 0.5% on the total energy generated	4–9% for coal
	-Catalyst lifetime: Coal from 6 to 12 years; oil from 8 to 12 years; natural gas: higher than 10 years	1–2% for the combined-cycle with natural gas
SNCR	-Reduction rates of 20–50%	
	-Electric energy consumption around 0.1–0.3% on the total energy generated	1–2%
	-Not suitable for gas turbines	

Source adapted from USEPA [16]

The typical efficiency of NO_x reduction for the SNCR technology can be 20–50% in stoichiometric injection levels (ammonia and urea) and in usual operating temperatures. However, the unevenness of temperature in the combustion system can limit the overall efficiency of NO_x reduction to values below 50%. Table 3 presents performance figures and characteristics of systems for the abatement of NO_x emissions of the End-of-Pipe systems.

In the case of turbines, due to the high temperatures involved (a condition that, in principle, maximizes the formation of NO_x) the techniques of prevention and abatement of emissions, that are partialy similar to those used in boilers will be treated as a specific item.

4.3 Nitrogen Oxides in Turbines: Techniques for Reducing Emissions and Their Impacts on Cost and Thermal Efficiency

In the last decade, the thermoelectric generation has expanded the use of gas turbines combined with steam turbines. Gas turbines provide higher thermal efficiency and can lead to significant emission reductions of nitrogen oxides.

Concerning NO_x formation, the thermal mechanism is responsible for the largest share of the NO_x formed. The portion of the NO_x produced from the nitrogen contained in the fuel, whereas natural gas is the fuel most widely used, can be considered as being negligible.

With regard to the formation of prompt NO, the portion formed by this mechanism may be important considering that the NO_x levels required for turbines in some situations are very low and that the levels of NO_x prompt produced are about 10 ppm. The values of prompt NO formed in turbines may be, in some situations, in close agreement with the emission standards.

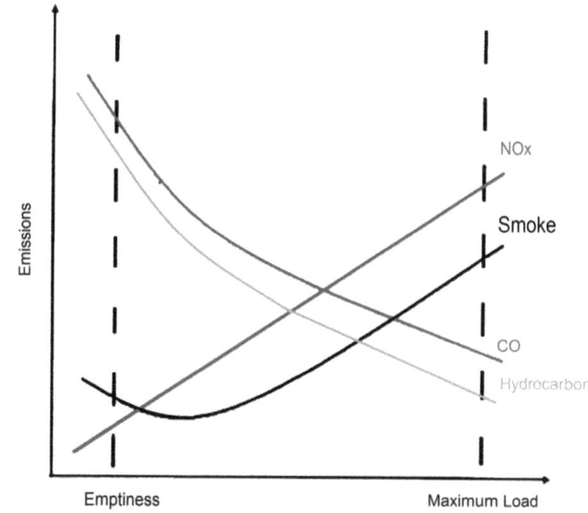

Fig. 5 Influence of the generated load in the rates of pollutant formation (*Source* adapted from Lefebvre [10])

As for the boilers, the mechanisms of formation of NO_x in turbines are strongly dependent on the process temperatures, nitrogen contained in the fuel, residence time, temperature of the intake air in the combustor, the inlet pressure in the combustor and the turbine load.

When the turbine operates at low loads, the formation of CO and hydrocarbons is preferred and the formation of nitrogen oxides is minimal. As power generation increases there is a reversal of that proportion, and when maximum load is reached, the generation of NO_x generation is maximum and CO and hydrocarbons reach minimal levels [10]. Figure 5 illustrates the influence of the charge generated in the emission rates.

4.4 Pollution Prevention Techniques for Controlling the Emission Levels of NO_x in Gas Turbines

The control of the emission rates of NO_x in turbines (and also in boilers) is performed employing techniques to prevent NO_x formation and End-of-Pipe. The prevention techniques can be divided into wet techniques, also known as Wet Low-NO_x and combustion control techniques or Dry Low-NO_x.

The wet control techniques include steam injection, while the dry, impoverished combustion, staged combustion and residence time reduction. The SCR and the $SCONO_X$ are End-of-Pipe control techniques and should, in most cases, be employed in combination with the prevention techniques.

Considering the fuel factor, equally to boilers, the flame temperature is directly related to the type of fuel burnt. Thus, fuels that present higher flame temperatures are likely to form more NO_x. Fuels such as syngas, that present flame temperatures

Table 4 NO_x emissions and efficiency and power increase as a function of the injected water ratio

NO_x (ppmv)	Fuel	Water/fuel ratio (%)	Power increase (%)	Efficieny increase (%)
75	Light oil	50 liquid	3	1.8
42	Natural gas	100 vapour	5	3
42	Natural gas	140 vapour	5	2
25	Natural gas	120 vapour	6	4
25	Natural gas	130 vapour	5.5	3

Source Gallego et al. [8]

lower than those of natural gas and liquified petroleum gas (LPG), for example, are preferred because, in principle, less NO_x will be formed.

4.5 Techniques of Wet Control: Wim Low-NO_x (WLN)

The control technique of NO_x by injecting water or steam in gas turbines reduces NO_x formation in approximately 70% but also reduces the efficiency of the thermodynamic cycle in 2–3%. However, the power output in the turbine shaft is increased by 5–6%, due to the increase of the mass flow through the turbine, which is proportional to the injection of steam or water [16]. Usually the injected water flow is around 50% of the fuel rate and, in the case of steam injection, the flow is 100–200% of the fuel flow.

Table 4 presents the NO_x emissions and efficiency and power increase as a function of the injected water ratio. Figure 6 shows the NO_x reduction rates for different rates of injection of steam and water for turbines operating with light oil and natural gas.

The analysis of the data presented in Table 4 and Fig. 6, indicate that the emission rate reduction is strongly dependent on the amount of water or steam injected. They also show that, to achieve high rates of reduction in the NO_x formation rates, large amounts of water or steam are required and in some cases the flow of the burnt fuel is outweighed.

Data presented by Rentz et al. [13] show that the highest reduction rates are obtained with water injection, given that the consumption of the energy of part of the fuel is intended to water vaporization, a procedure that helps to reduce temperature and therefore, the thermal NO_x formation rate is reduced. The data also show that, in order to achieve the same rates of reduction of NO_x, approximately twice as much injected liquid water is required when compared to steam injection.

The injection of liquid water is often used when there is no availability of steam, for example, in simple cycle applications. However, steam injection should be prioritized because it reduces the occurrence of small cracks resulting from thermal shock. Regardless of being water or steam, the rate control of the NO_x

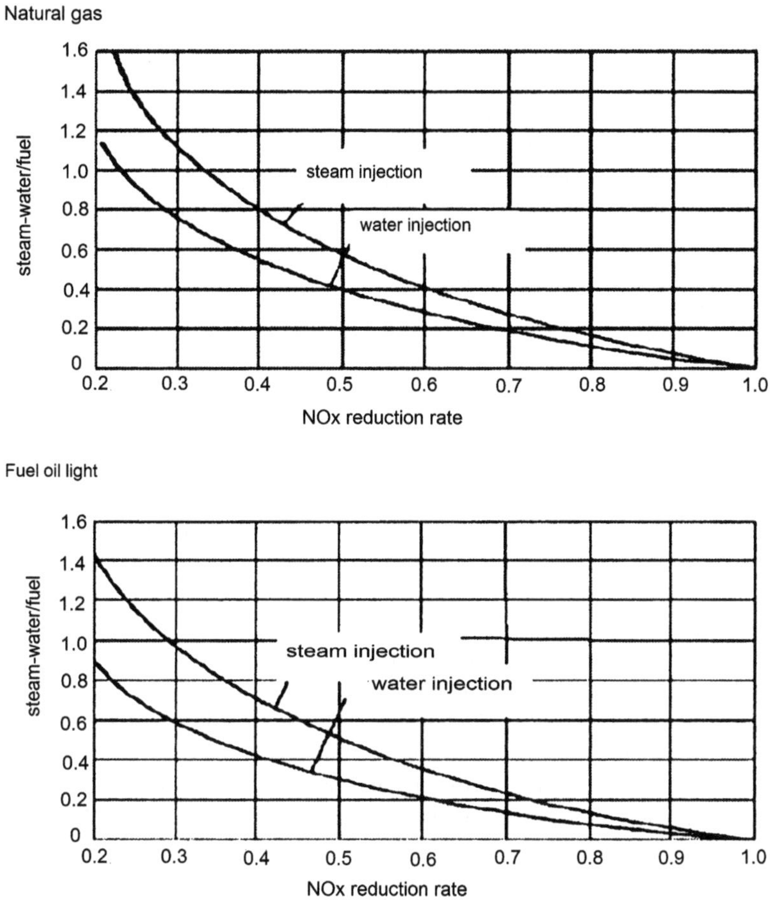

Fig. 6 No_x reduction rates for different rates of injection of steam and water for turbines operating with light oil and natural gas. (*Source* adapted from Rentz et al. [13])

formation by the wet technique shortens turbine life due to thermal shock and increases the risk of failure [13].

4.6 Dry Control Techniques: Dry Low-NO_x (DLN)

The use of the technique of controlling NO_x emissions by the Dry Low-NO_x method is widespread in large capacity thermal power plants. Recently this technique has been extended and applied to units with a 20 MWe capacity.

The use of the Dry Low-NO_x techniques is intended to reduce the temperature and the residence time that are important factors to reduce the rate of formation of NO_x.

The DLN technique employs the reduction of the pre-mix fuel–air relation to decrease the NO_x formation rate. Besides the reduction of the flame temperature peaks it provides a most efficient temperature distribution along the flame. This reduction technique is employed along with the combustion air and the fuel staging technique to reduce the residence time of the gases in regions where temperature peaks occur inside the combustion chamber.

For the staging technique, the combustion chamber can be considered as being divided into two zones. In the first zone the combustion takes place with a deficiency of air. The consequences are the existence of low temperatures and CO and hydrocarbons are produced. In the second zone, the combustion of CO and the hydrocarbons formed in the first zone occurs with the excess of air that limits the flame temperature. Both techniques (the impoverished combustion and combustion staging) when employed, limit the NO_x formation rate by the thermal mechanism. Depending on the combustor design, the levels of the NO_x formation rates are from 9 to 25 ppmv.

However, despite of significantly reduction on the NO_x formation rate, the techniques of staged combustion and impoverished combustion and can cause instabilities in the flame and raise the rates of CO formation. To minimize these effects, pilot burners are commonly installed to keep the stability of the flames.

Capital costs are a function of the facility size and also of the characteristics of the turbines installed. The increase in capital cost in relation to systems that have no control of the NO_x emissions is around 15%. The operating cost is estimated at 40% As an example, in 2000 NESCAUM estimated that for turbines with capacities greater than 75 MW, the approximate cost was U.S. $ 1,000/ton of NO_x reduced for all the operating conditions and only a few hundred dollars per ton of NO_x reduced when the turbine operates at a capacity factor of 0.85. Another strand of the development in the field of NO_x formation rate control is the catalytic combustion technique in turbines. The technique is still undergoing improvement.

Catalytic combustion is applicable for pre-combustion mixture chambers with lean burn fuel. This process reduces NO_x emissions to values as low as 3 ppm by burning the fuel in the presence of a "flameless" catalyst. Low levels of NO_x are obtained while the levels of CO and unburned hydrocarbons do not exceed 10 ppm.

In catalytic combustion, the process occurs in two steps. In the first stage, a partial combustion takes place inside the catalyst at a controlled temperature which virtually precludes the formation of NO_x. The second stage of the process goes downstream with a "flameless" homogeneous type reaction that facilitates the formation of virtually residual amounts of NO_x.

4.7 End-of-Pipe Techniques for Controlling the Levels of NO_x Emissions in Gas Turbines

The techniques Selective Catalytic Reduction (SCR) and End-of-Pipe SCONOx are applied for the abatement of the NO_x emissions in turbines. The SCR systems employ ammonia and the reactions take place on the surface of the catalyst at high temperatures. The reaction of NO_x with ammonia leads to the formation of nitrogen and water.

In a simple configuration of an SCR system for NO_x abatement in the exhaust gases of a turbine, the hot gases present flow to the heat recovering systems where the catalyst bed is installed. The reaction of ammonia with NO_x within a certain range of temperature takes place at the bed surface. Ammonia is injected uniformly into the gas flow, immediately before the catalytic bed. The operating temperature of commercial catalysts should not exceed 400°C.

The work range of the catalyst requires that the thermodynamic arrangement of a power plant that employs an SCR system for NO_x abatement has to be different from a simple thermodynamic cycle. That requirement is due to the fact that the exhaust gases reach temperatures greater than the limit imposed by the catalyst. Thus an arrangement that includes cogeneration is suitable for the SCR systems.

An excess of ammonia emissions (called ammonia slip) is generated in turbine operations, because the process efficiency is lower than 100%.

A common issue to any post-combustion system is the increase of the restriction imposed to the flow of exhaust gases, which results in increased consumption of fuel by the turbine in order to maintain the output load.

The use of the SCONOx technique for the abatement of NO_x emissions is relatively new and poorly disseminated. According to some manufacturers, as the technique is employed, the NO_x emissions are around 2 ppm of NO_x and 1 ppm of CO [16].

The process takes place in two steps. In the first, NO is oxidized to NO_2 and CO to CO_2. The NO_2 formed is then adsorbed on the surface of the catalyst which is periodically regenerated. In the regeneration process, NO_2 is converted into N_2 and hydrogen (H_2) and O_2 into water. This technology has the advantage of not using ammonia as a reagent and therefore its emission is avoided. Table 5 presents a comparison of the emission levels of the technologies used to reduce NO_x emissions for natural gas and light oil for the exhaust gases generated in turbines.

Table 6 shows a comparison of control costs of NO_x emissions as a function of the load generated and the control technology applied to the exhaust gases from gas turbines.

Table 7 presents a comparison among several control costs for several NO_x control technologies applied to gases generated in gas turbines.

5 Particulate Matter: Impacts on Cost and Thermal Efficiency and Technologies Applicable to Emission Reduction

Power plants, mainly those that burn coal, fuel oil, petroleum coke and biomass release particulate matter in the exhaust gases. Among all impacts on health resulting from thermal power generation, the particulate material released to the

Table 5 NO_x emissions levels in gas turbines according to the technology for reducing NO_x and fuel

NO_x control techniques	Natural gas (ppm)	Light oil (ppm)
Uncontrolled emissions	155	240
Wet techniques	25	42
Dry techniques	9	42
Selective catalytic reduction	2–5	4–10
Catalytic combustion	3	Not applicable
SCO NO_x	1–3	Not applicable

Source adapted from USEPA [16]

Table 6 Comparison of control costs of NO_x emissions as a function of the load generated and the control technology applied to the exhaust gases from gas turbines

Turbine power output	5 MW class		25 MW class		150 MW class	
NO_x emission control technology	$/ton	¢/kWh	$/ton	¢/kWh	$/ton	¢/kWh
DLN (25 ppm)	260	0.075	210	0.124	122[a]	0.054[a]
Catalytic combustion (3 ppm)	957	0.317	692	0.215	371	0.146
Water or steam injection (42 ppm)	1.652	0.410	984	0.240	476	0.152
Conventional SCR (9 ppm)	6.274	0.469	3.541	0.204	1.938	0.117
High temperature SCR (9 ppm)	7.148	0.530	3.841	0.221	2.359	0.134
Low temperature SCR (9 ppm)	5.894	1.060	2.202	0.429		
SCONO$_x$ (2 ppm)	16.327	0.847	11.554	0.462	6.938	0.289

[a] 9–25 ppm
"¢/kWh" based on 8,000 load hours per year
Source adapted from DOE [5]

atmosphere is considered as being the most significant. The magnitude of damage due to inhalation of particles by the population depends mainly on the chemical and physical properties of particles emitted. The size and also the density, influence the depth degree that the particles penetrate the respiratory system, i.e, they define the capacity in which the particle will reach the lungs and which region is most affected (upper or lower regions of the respiratory system). The chemical composition, in turn, determines the toxicity and risks to which the population is exposed due to particulate matter inhaled. Several epidemiological studies have shown that there is a strong correlation between the standards of air quality (for total suspended particles), with morbidities such as asthma, bronchitis, emphysema, heart disease, etc. There are also studies that indicate that diseases are also associated with very small diameter particles, less than 2.5 μm (MP 2.5).

5.1 Particulate Matter: Formation and Abatement Techniques

The particulate matter generated in the combustion processes are classified in two basic types: cenospheres (coke) and soot. The fraction of cenosphere called coke is

Table 7 Comparison of several control costs for several NO_x control technologies applied to gases generated in gas turbines

Control technology	Turbine output (MW)	Emission reduction (ppm)	USD/t	USD cents/kWh
Water/steam injection	4–5	Not controlled. – 42	1,500–1,900	0.39–0.43
Dry Low-NO_x	4–5	Not controlled. – 42	NA[b]	NA[b]
Dry Low-NO_x	4–5	Not controlled. – 25	270–300	0.006–0.09
Catalytic combustion	4–5	Not controlled – 3	1.000	0.32
SCR–low temperature	4–5	42 – 9	5.900	1.06
lSCR–conventional	4–5	42 – 9	6.300	0.47
SCR–high temperature	4–5	42 – 9	7.100	0.53
$SCONO_x$	4–5	25 – 2	16.300	0.85
Water/steam injection	20–25	Not controlled. – 42	980	0.24
Dry Low-NO_x	20–25	Not controlled. – 25	210	0.12
Catalytic combustion	20–25	Not controlled. – 3	690	0.22
SCR–low temperature	20–25	42 – 9	2.200	0.43
lSCR–conventional	20–25	42 – 9	3.500	0.2
SCR–high temperature	20–25	42 – 9	3.800	0.22
SCONOx	20–25	25 – 2	11.500[3]	0.46
Water/steam injection	160	Not controlled. – 42	480	0.15[c]
Dry Low-NO_x	170	Not controlled. – 42	124	0.05[c]
Dry Low-NO_x	170	Not controlled. – 25	120	0.055[c]
Catalytic combustion[a]	170	Not controlled. – 3	371	0.15[c]
lSCR–conventional	170	42 – 9	1.940	0.12[c]
SCR–high temperature	170	42 – 9	2.400	0.13[c]
$SCONO_x$	170	25 – 2	6.900	0.29[c]

[a] Plant operation started in 1999. Annual cost provided by the manufacturer
[b] 'NA': not available ou obsolete technology in 1999
[c] Based on the costs of a 83 MW power plant. Size has been scaled-up for the desired plant
Source adapted from IPCC [9]

made of much larger particles (1 μm up to 100 μm) than soot. They are derived from fuel ash and coke material. Soot particles are small with diameters less than 0.1 mm of polymeric nature resulting from cracking reactions followed by the polymerization of hydrocarbons in vapor phase, generating condensed nuclei.

The formation of coke depends on the characteristics of the combustion system and operating conditions, but is also strongly dependent on the specifications of the fuel. In the specific case of fuel oil, the formation of cenosphere is strongly dependent on the nature and the characteristics of petroleum refining, the ash content and the content of compounds called asphaltenes, which are long-chain hydrocarbons, difficult to disintegrate and that contribute greatly to the formation of the coke particles. For solid fuels such as coal, shale, peat, etc., the particulate matter generated is almost entirely derived from the ashes and not consumed fuel particles entrained with the gases called fly ash. A very small portion may be derived from the volatilization and polymerization of the volatile portion existing in the fuel. The amount of total particulate matter generated by the solid fuel use is, in some extent, dependent on the ash content of fuel and the type of combustion equipment used.

The generation of particulate matter resulting from the use of gaseous fuels can be considered negligible if the combustion conditions such as the mixture of fuel and combustion air, the amount of air available for the combustion and the residence time in the gas phase at high temperatures are adequate. However, considering that soot is formed from volatized hydrocarbons, it is possible that the formation of reasonable amounts in gas flames occur. Despite of presenting a small size when compared to the chemosphere and inorganic particulate material (even though the amount of soot mass is negligible in number of particles) their amount can be significant to the extent that visual impact is quite significant. It is noteworthy that, if on one hand the presence of soot in the chimney is detrimental from the standpoint of environmental issues and public health, in the interior of the combustion chamber when the thermal exchanges by radiation are important, on the other hand soot plays a fundamental role in the behavior of flame and its interaction with the environment. In many situations, purposely there are areas that promote its formation and posterior ones where the soot formed is destroyed.

The techniques for reducing the emissions of total particulate matter can be prevention techniques and End-of-pipe that abate part of the emissions generated. Fuel switching and the use of good combustion procedures are techniques for preventing or reducing the generation of particulate matter. They involve the control of the air excess, temperature inside the chamber, residence time and the mixing of the fuel with combustion air (turbulence).

The replacement of high ash content coals by other fuels that present lower ash contents such as fuel oil or gas is a prevention technique that reduces the amount of the particulate matter generated proportionally to the amount of reduced ash.

As previously mentioned, the use of gaseous fuels such as natural gas and liquefied petroleum gas (LPG), is a procedure to prevent particulate matter emission, because the generation of ash and cenospheres is avoided. However, if the combustion parameters are not adequately controlled, the gaseous state is conducive to the formation of soot.

However, in many cases of retrofit in power plants, the replacement of fuel with a potential for generating sooty flames by another of lower potential, can cause loss of thermal efficiency due to reduced rates of heat exchange by radiation in higher temperature zones.

Considering the prevention techniques, an analysis of the mechanisms of formation of both, soot and cenospheres, indicate that a good rate of mixing between fuel and oxidant (oxygen), proper temperature in the combustion chamber, adequate residence time for the higher temperature species as well as adequate amounts of available oxygen available, are necessary conditions for reducing the formation of particulate matter.

Electrostatic filters and fabric filters are end-of-pipe techniques employed to reduce the emission rates of particulate matter. The choice of one type or another

Table 8 Characteristics of control systems for emissions of particulate matter

Control device	Characteristics
Electrostatic filters	• Estimates of removal efficiency for particles smaller than 1 micron are greater than 96.5%, while for particles larger than 10 μm the efficiency is greater than 99.5%
	• Power consumption is estimated as 0.1–1.8% of the electricity generated
	• There is the need to adjust the resistivity of the gas, because it has restrictions for particles with high electrical resistivity
	• Features low pressure drop even at high flows
Fabric filters	Estimates of removal efficiency for particles smaller than 1 micron are greater than 96.6%, while for particles larger than 10 μm, the efficiency is greater than 99.5%
	Power consumption is estimated as 0.2–3% of the electricity generated
	The life of the fabric sleeves is a function of the chemical characteristics of the gas; so for gases with high sulfur levels, life is reduced
	Operating costs are affected by the characteristics of the gas (because it plays an influence on the fabric sleeves) and by the particle size. The size of the particles to be retained determines the type of mesh structure, which is an important factor in composition of capital and operating costs

Source World Bank Group [17]

depends mainly on the characteristics of the fuel used, the systems used to control emissions and other pollutants and especially, emission standards and the physical characteristics of the generated particles.

Thus, for processes where the generation of soot is high, the use offabric filters must be prioritized because of its high collection efficiency for smaller particles. Moreover, the use of fabric filters has a strong restriction concerning the gas temperature that should not exceed the temperature that the fabric sleeve supports. In addition to technical aspects, capital and operating costs must be considered for the definition of the abatement device. Table 8 presents explanation of the most efficient technologies (Best Available Technology—BAT) to control emissions of particulate matter generated in power plants and considerations for reducing thermal efficiency of cycles due to the systems of abatement of particulate matter. Cost information and operational costs of capital are dependent on the configuration of systems and fuels burnt.

6 Carbon Dioxide: Techniques to Reduce Emissions and Their Impacts on Cost and Thermal Efficiency

The control of the emissions of greenhouse gases is a mandatory factor to the mitigation of CO_2 emissions in power plants as it is for other industrial processes that use fossil fuels.

Carbon dioxide is the main compound resulting from the complete combustion of fossil fuels and other fuels that contain the element carbon. It can also result from the anaerobic respiration of living beings.

CO_2 is responsible for about 64% of the greenhouse effect. Around 6 billion tonnes of CO_2 are thrown into the atmosphere every day. The residence time of CO_2 in the environment is referred from 50 to 200 years. Carbon monoxide (CO) results from the incomplete combustion of fossil fuels and other materials containing the element carbon in their composition. It is removed from the environment mainly by the soil where it is converted to CO_2. The residence time of CO in the environment is from 1 month to 2.7 years.

In addition to the alternative use of diverse sources of electric energy production (nuclear, hydro, biomass, solar, wind) to replace fossil fuels, there are alternatives that seek to mitigate the CO_2 that is already being delivered by existing thermal plants and those which will be constructed on the basis of non-renewable fuels.

In this case, the following alternatives are being considered for the mitigation of CO_2 emissions: the efficience increase of converting thermal energy into electricity:

- the replacement or use of fuels with lower carbon contents;
- the implementation of systems for capture, storage for CO_2 sequestration. (CO_2 Capture and Storage, CCS).

Within the concept of cleaner production, improvements on the efficiency of thermal cycles are a desirable goal, since fuel consumption decreases per unit of electricity generated and, therefore, there is a proportional decrease of the CO_2/kWh emissions.

Simbeck and Roekpooritat [15] estimate that it is possible to achieve overall gains on efficiency in coal power facilities of about 2% from the present average values (33–35%), at moderate cost with a change of equipment for the improvement of steam quality, with changes in temperature levels and pressure of the cycle. Modifications to the steam generators and turbines could result in the reduction of CO_2 emissions of around 6%, still considerable below the global targets set.

In combined CHP (Combined Heat Power-CHP) the efficiencies of global cycles can be significant, from 33 to 80%. The effectiveness of the deployment of such systems is subject to the system, load, installment and heat to be compatible, or in some way, synchronized with the share of electricity generated. Thus, this measure has been quite feasible for industrial plants where there are greater possibilities for thermal power use and less on the basis that generally meet the demand for electricity. In a general manner, the higher the carbon/hydrogen ratio of the fuel, the greater will be the specific emission of CO_2/kWh generated. Under this standpoint, among fossil fuels, an increasing scale of CO_2 emissions can be observed according to an order of the most environmentally friendly to the least favorable: natural gas, light fuel oil, residual fuel oil, lignite, sub- bituminous, bituminous, anthracite and coke coals. The specific emissions of CO_2 are associated not only to the chemical composition of the fuels, but also to the efficiency of converting thermal energy into electrical energy and these, in turn, are conditioned by the technologies employed in the thermal power generation cycles adopted.

There are several comparative advantages of natural gas compared to other fossil fuels with emphasis on the environment, since the specific CO_2 emissions can be of up to three times smaller than those resulting from coal combustion.

Despite the quite favorable numbers of replacing coal by natural gas and the low costs involved in the conversion, Simbeck and Roekpooritat [15] observe that there is no sufficient guarantee of long-term supply of natural gas. The coal plants are responsible for 51% of electric generation in the USA, and the replacement of coal by gas, would represent a very significant additional demand, with very significant impacts on the gas prices and therefore, proportional increases in tariffs electricity. However, according to data compiled by the DOE [6] in the U.S., the specific cost per unit of energy for natural gas and oil in 2008 were respectively 1,100 and 900 Cents/MMBtu, about 3.7–3 times the values observed in 1997. In the same period, the average costs of fossil fuels have evolved from 150–400 Cents/MMBtu, which represents a smaller increase for gas, 2.7 times, and therefore less impact on operating costs.

In average values observed in Brazil, where biomass is used for energy purposes, significant differences among the same indices are observed. The values of natural gas and residual fuel oil, both used in thermal power generation, are currently 4.3–5.8 times the respective values observed for wood or coal, or 3.6–4.8 times the same content for sugarcane bagasse.

Fuel blending is an already common practice in coal power plants and it is motivated not only by the need to reduce operating costs, but also, by environmental issues such as mitigation measures for CO_2 emissions. The blending of fuels of different qualities that results in the reduction of mean C/H ratio, usually results on the decrease of CO_2 emissions and also, NO_x, SO_2, mercury and particulate matter emissions.

Miller and Tilman [12] describe a typical case of an American large thermoelectric power plant (3,100 MWe), where the management of the process of blending fuel is guided primarily by controlling atmospheric emissions.

Motivated by the increased in the global availability of coal compared to other fossil fuels, technological development moves into the direction to make it more environmentally adequate, avoiding the release of CO_2 to the atmosphere.

According Rubin and Rao [14], in USA, the CO_2 emissions from coal-fired power plants accounted for 79% of emissions to the atmosphere exclusively for electric power generation, although its share in generation has been only 51%. Even with the expected growth of natural gas for 2020, which will result in reducing the share of coal predicted as 44%, in absolute values, the consumption of coal will increase (relatively to values of the period).

Cleaner technologies to burn coal known as "Clean Coal Combustion Technologies," to separate, capture and store CO_2 (CCS) has been investigated, particularly in countries where the participation of coal power is a major basis for the supply of electric power supply. That is the case of Canada and USA.

While ensuring the effective sequestration is still a subject of inquiry, the use of systems for this purpose have been considered for both, the existing conventional power plants and for the new facilities which employ alternative technologies that have already incorporated the technique into their design and project.

In conventional combustion systems, air and CO_2 concentration in the effluent gases are relatively low, 10–12% (volume basis). It means a difficulty due to the large volumes of inert gas (N_2) to be treated, implying in high costs of separation and capture. In these cases, the CO_2 capture systems frequently used are based on the absorption in amine solutions, as well as on technical solutions that have already been commercially developed and disseminated. The gas stream passes through scrubbers through which the amine solution circulates. CO_2 is extracted from the gas flow. The amine solution is regenerated in another vessel by the use of reasonable amounts of heat and electricity for powering pumps and other peripheral equipment. Deducted from the electricity generated, the overall net cycle efficiency is decreased.

While ensuring the effective sequestration is still a subject of inquiry, the use of systems for this purpose have been considered for both, the existing conventional power plants and for the new facilities which employ alternative technologies that have already incorporated the technique into their design and project.

Other alternative separation techniques consider the increase of the CO_2 concentration in the stream of combustion products as a way to reduce the volume to be treated, and therefore the capital and operating costs (which ultimately reduce the availability of electric power by penalizing overall efficiency).

Technological development has advanced greatly in the recent years in view of solving the conflict between reducing the environmental impact and penalizing the efficiency of conversion of fuel into net electrical energy. Different techniques for capturing and sequestering CO_2 were subject to execution of several research studies at the level of experimentation as well as other studies that resulted in the design and construction of several large units with different arrangements.

The oxy-fuel combustion that promotes pure O_2 instead of air, significantly reduces the volume of combustion products and increases the concentration of CO_2 in the gases to be treated, improving the conditions for their separation and capture and results in lower capital costs.

A technique under researched and that has not been applied in full-scale in combustion chambers of boilers (that operate at pressures close to the atmosphere) is the use of combustion air enriched with oxygen or pure oxygen ($\sim$100%) combined with gas recirculation with high levels of CO_2.

According to Beer [1] studies show that in coal flames, as O_2 and CO_2 are employed to substitute air, significant reductions of NO are observed and the resulting effuent gases present CO_2 levels above 98% (by volume, dry basis). They can be removed in systems with lower gas volumes when compared to the conventional air systems.

Another alternative scheme in boilers burning hydrocarbons proposes the use of pure oxygen associated to water injection with the water vapor recirculation. The CO_2 generated in the combustion (10%) is separated from the gases by water condensation (90%) in steam turbines which must operate at higher temperature ranges than in conventional turbines. It has been estimated that such modification on the Rankine cycle can achieve a calculated efficienct of 63%, higher than the 50% possible in conventional combined cycles, for a net power generated of around 100 MWe.

The gasification of solid fuels can be considered an incomplete combustion process, where part of the fuel is oxidized just to heat the solids and trigger the processes of pyrolysis and cracking of the hydrocarbons present, producing a fuel gas.

The gasification process, when it occurs at pressures (up to 60 bar) and high temperatures ($\sim$1,300°C) in the presence of O_2 and water vapor, produces combustible gases of high calorific value and appreciable amounts of CO, H_2 and CH_4 ("syngas"). The fraction of CO_2 generated in the combustion of the fuel is separated from other components of the gas fuel by special catalytic separation systems.

The integration of the pressurized gasifiers with combined cycle gas turbines with steam turbines has been considered in several studies, as a promising alternative to reduce the environmental impacts of using coal in power generation, especially by enabling CO_2 sequestration with lower power consumption compared to other alternatives.

A recent report issued by DOE [5] presents a comparative analysis between the different technologies already available and tested. Twelve different configurations of power plants with installed power rating of around 550 MWe were simulated. Among the 12 simulated scenarios, six possibilities for separation and capture of CO_2 generated in the combustion of coal and natural gas were considered.

The comparative analysis presented in Table 9 considers three basic configurations of combined cycle, namely: Combined Natural Gas (NGCC—Natural Gas Combined Cycle); two Combined Cycles with the Combustion of Pulverized Coal (subcritical and supercritical boilers) to (PCB—Pulverized Coal Boiler) and three cycles combined with the coal gasifiers Integrated (IGCC—Integrated Gasification Combined Cycle) with gasification technologies developed by different manufacturers: GHG—General Electric Energy, Conoco Phillips—CoP and Shell.

The advantages of the use of natural gas compared to coal in all aspects presented in the Table 9 are well-known. The specific emissions of CO_2 up to three times higher for coal in relation to corresponding natural gas are due to two related factors: the higher the C/H ratio in the coals (for the alternative PCC) the lowest overall efficiency of the cycles due to the high energy consumption in the gasification step (to produce O_2) and CO_2 capture and storage.

The situation is very favorable to coal emissions in settings that have systems for capturing and storing CO_2, when compared with natural gas conventional installations without CO_2 separation and storage systems. In this aspect, the configuration with gasifiers (IGCC), present major advantages as compared to the former, since it uses oxygen and steam as gasification agents, which provides the best conditions for the separation and capture of CO_2.

Considering capital costs, coal-fired units have costs of up to 2.5 deployment costs for units with natural gas. However, economic feasibility studies which take

Table 9 CO_2 emissions, cost and overall efficiency provided for the various cycles of power from fossil fuels

		IGCC (Integrated gasification combined cycle)	PCC (Pulverized coal boiler)	NGCC (Natural gas combined cycle)
CO_2 emission factor (kg/MWh)	Without CCS*1	752–796	804–855	361
	With CCS*1	90–114	115–126	42
Cycle Global Efficency (%)	Without CCS*1	38.2–41.1	36.8–391	50.8
	With CCS*1	31.7–32.5	24.9–27.2	43.7
Capital cost (US$/kW)	Without CCS*1	1.733–1.977	1.549–1,575	554
	With CCS*1	2.390–2.668	2.870–2,895	1.172

(*1) CCS-CO_2 captured and storage
Source DOE [5]

into consideration all other aspects including the regional availability of fuel, water and other inputs, fuel costs and electricity tariffs, circumstantially may offset the higher capital costs and the lower conversion efficiency.

7 Dual Burn

The technological developments in thermoelectric generation cycles observed in the late nineteenth century and the beginning of century XX are graded in increasing the overall efficiency and reducing emissions of pollutants. It resulted in the proposition of different arrangements and schemes involving more than one fuel in the same generation or configuration arrangement.

Among the various possibilities for models of thermal power generation cycles studied, the schemes that combine gas turbines (Brayton at 1,660–900 K) with steam turbines (Rankine at 850–288 K), enable the achievement of higher values of thermal efficiency compared to conventional steam cycles, because the operation temperature ranges are complementary.

At the current stage of development, in the cycles that use only natural gas ("GCC—Natural Gas Fired Gas Turbine-Steam Combined"), the overall efficiency is located around 60%. Regarding the impacts on air emissions, according to Beer [1], these can be considered to have the least impact among the alternatives that use fossil fuels, whose major compounds emitted pollutants are CO_2 (345 g/kWh) and NO_x (355 mg/kWh).

Considering the perspectives of natural gas price rising resulting from uncertainties about its actual availability, the higher availability of coal and the possibilities of using biomass, other configurations have been adopted, some using more than one type of fuel (liquid or solid). They are generically called combined burn ("dual Fuelled cycles") being suitable for gas turbines and boilers.

Even the combustion chambers of turbines designed to burn natural gas show some flexibility and gaseous and liquid fuels can be switched. Considering the design of the turbines, the liquid fuels must present low viscosity and tighter specifications concerning the presence of specific contaminants (ash, sulfur, etc.). Otherwise the integrity of the turbine will be committed leading to an increase of maintenance costs.

Pilot experiments using double-fluid nozzle type ("dual fuel nozzles") to spray light liquids in turbines built by GE (originally gas turbines) led to good performance and significant operating cost. Recently In Brazil, some experiments were accomplished in order to substitute natural gas by ethanol. The results were considered as being satisfactory and the procedure will be implemented in several units in the near future.

In the case of the cycles with gasifiers with CO_2 separation, the process of fuel decarbonation takes place by the conversion of the CO produced into CO_2 through the "shift" reaction ($CO + H_2OCO_2 + H_2$) increasing the concentration of H_2 in the fuel gas. Subsequently, CO_2 and sulfur compounds are extracted from the fuel gas ("syngas") before being introduced into the combustion chamber of the turbine.

In turbine operation, in addition to the impacts of the fuel injection systems arising from the change in the composition of the fuel gas, it necessary to replace the air as a diluent of the combustion gases. In these cases, steam is employed by direct contact of the fuel gas (syngas) with water or with nitrogen from the Air Separation Units (ASU). Nitrogen is produced from the oxygen used as the gasification agent of solid fuel.

The installation of the generation and processing systems of fuel gas prior to its use in turbines present high capital and operational costs and also, represent an increase on electrical energy consumption. This means a reduction of the net availability of the cycle resulting in the values presented in Table 7.

In the combustion chambers of boilers the use of more than one type of fuel simultaneously or alternately is a fairly common practice, particularly in conventional boilers (frontal or tangential burning) In the case of solid fuels the feed is usually made in the powder form.

The technique of reburning with natural gas as a strategy for the control of the NO_x emissions in the combustion of waste oil or pulverized coal, can be considered as a combined burning technique.

In boilers designed for fluidized bed combustion, the flexibilities concerning the use of different fuels with respect to their physical state (gas, liquid and solid) are smaller.

References

1. Beer JM (2000) Combustion technology developments in power generation in response to environmental challenges. Progr Energ Combust Sci 26:301–327
2. Blauwers J, Smets B, Peeters J (1977) Mechanism of prompt no formation in hydrocarbon flames. In: Proceedings of the 17th symposium (international) on combustion, pp. 1055–1071
3. Bowman CT (1973) Kinetics of nitric oxide formation in combustion processes. In: proceedings of the 14th symposium (international) on combustion, pp 729–738

4. Carvalho JA, Júnior PTL (2003) Emissões em processos de combustão. Editora UNESP, São Paulo (in Portuguese)
5. DOE—U. S. Department of Energy/NETL—National Energy Technology Laboratory (2010) Cost and performance baseline for fossil energy plants, volume I: bituminous coal and natural gas to electricity—final report. MIT Coal Retrofit Symposium
6. DOE—U. S. Department of Energy/NETL—National Energy Technology Laboratory (2010) Cost and performance baseline for fossil energy plants: volume I: bituminous coal and natural gas to electricity revision 2. DOE/NETL-2010/1397
7. Fenimore CP (1971) Formation of nitric oxide in premixed hydrocarbon fames. 13th symposium (international) on combustion, The Combustion Institute, pp 373–379
8. Gallego AG, Martins G, Gallo WLR (2000) Emissões de NO_x em turbinas a gás: mecanismos de formação e algumas tecnologias de redução. Rev Ciênc Tecnol 15:13–22 (in Portuguese)
9. IPCC (2006) Integrated pollution prevention and control reference document on best available techniques for large combustion plants. European Commission
10. Lefebvre AH (1983) Gas turbine theory. McGraw-Hill, New York
11. Miller JA, Bowman CT (1989) Mechanism and modeling of nitrogen chemistry in combustion. Progr Energ Combust Sci 15:287–338
12. Miller BG, Tillman DA (2008) Combustion engineering issues for solid fuel systems. Academic Press, New York
13. Rentz O, Nunge S, Laforsch M (1999) Technical background document for the actualisation and assessment of UN/ECE protocols related to the abatement of the transboundary transport of nitrogen oxides from stationary sources. In: Integrated pollution prevention and control reference document on best available techniques for large combustion plants, Europe Commission
14. Rubin ES, Rao AB (2002) A technical, economic and environmental assessment of amine-based CO_2 capture technology for power plant greenhouse gas control. Washington, DC. doi:DOE/DE-FC26-00NT40935
15. Simbeck D, Roekpooritat W (2009) Near-term technologies for retrofit co_2 capture and storage of existing coal-fired power plants in the United States. MIT Coal Retrofit Symposium, 2009
16. USEPA (2000) Control of Nitrogen Oxides Emissions. APTI Course 418. Environmental Research Center Research Triangle Park, NC Air Pollution Training Institute (APTI)
17. World Bank Group (2008) Environmental, health, and safety guidelines—thermal power plants
18. Zeldovich YB, Sadovnikov PY, Frank-Kamenetskii DA (1947) Oxidation of nitrogen in combustion. Academy of Sciences, Moscow

Fundamentals of Reliability

A. P. Teixeira and C. Guedes Soares

Abstract This chapter aims at introducing the basic concepts and techniques relevant for reliability and availability analysis of systems. It starts by introducing the basic reliability concepts and the most important reliability models. The techniques commonly used for evaluating the reliability of systems with different configurations including series, parallel, series–parallel, and *k*-out-of-*n*, standby and more complex systems are then addressed. Finally, the availability concept and the reliability considerations for repairable components and systems are also addressed.

1 Introduction

This chapter on "Fundamentals of Reliability" briefly introduces the main and elementary reliability concepts that provide the necessary foundations for more specialized studies in the different application areas of reliability engineering. The selected topics cover the basic reliability concepts, the most important reliability models that quantify the reliability of components, the techniques commonly used for evaluating the reliability of systems with different configurations and the reliability considerations for repairable components and systems.

Although the present text presents an overview on the various topics, it does not provide an extensive and detailed description of the subject matter. The level of detail on each of those concepts is conditioned by space limitations, but references are provided to some textbooks that have similar approaches to the treatment of this subject.

A. P. Teixeira · C. Guedes Soares (✉)
Centre for Marine Technology and Engineering (CENTEC),
Technical University of Lisbon, Instituto Superior Técnico, Lisbon, Portugal
e-mail: guedess@mar.ist.utl.pt

G. F. M. de Souza (ed.), *Thermal Power Plant Performance Analysis*,
Springer Series in Reliability Engineering, DOI: 10.1007/978-1-4471-2309-5_5,

2 Basic Reliability Concepts

2.1 Reliability Function and Cumulative Distribution of Time to Failure

In general terms *reliability* is 'the ability of an entity to perform a required function under given conditions for a given period of time'. In technical terms, reliability is measured by the probability that a system or a component will work without failure during a specified time interval under given operating conditions.

Often the specified time interval or, alternatively, the mission duration is considered as a parameter t, and the probability $P[T > t]$ that the random variable time to failure T, will be greater than t is given by the reliability function $R(t)$, also referred to as the survival function,

$$R(t) = P[T > t] \tag{1}$$

The reliability function expresses the probability that the system operates without failure in a time period t, and therefore, since a system that does not fail for $T \leq t$ must fail at some $T > t$, then:

$$R(t) = P[T > t] = 1 - F(t) = 1 - P[T \leq t] \tag{2}$$

where $F(t)$ is the cumulative distribution function (CDF) of the time to failure that gives the probability, $P[T \leq t]$, that the time to failure T will be smaller or equal than the specified time t, or in other words, the probability that the system will fail at a time less than or equal to t.

The cumulative distribution function $F(t)$ and the reliability function $R(t)$ are related to the probability density function (pdf) $f(t)$ of the time to failure by:

$$f(t) = \frac{d}{dt}F(t) \tag{3}$$

$$f(t) = -\frac{d}{dt}R(t) \tag{4}$$

where $f(t)$ describes how the failure probability is spread over time. Basic properties of the probability density of the time to failure are: (1) $f(t)$ is always non-negative and (2) the total area beneath $f(t)$ is always equal to one. In the infinitesimal interval $]t,\ t + dt]$, the probability of failure is $f(t)\,dt$. The probability of failure in any specified time interval $t_1 < T \leq t_2$ is therefore given by:

$$P[t_1 < T \leq t_2] = \int_{t_1}^{t_2} f(t)\,dt = F(t_2) - F(t_1) \tag{5}$$

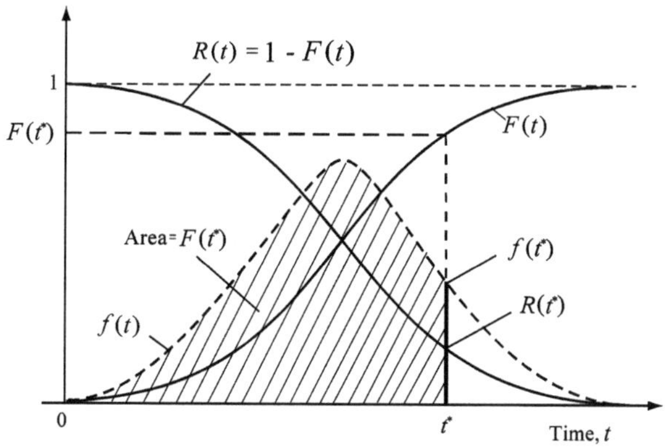

Fig. 1 Reliability function, cumulative distribution function of the time to failure and failure density function

Figure 1 illustrates the link between the reliability function $R(t)$, the cumulative distribution function $F(t)$ and the probability density function $f(t)$ of the random variable time to failure T.

An important reliability measure is the mean time to failure (MTTF), which is just the expected or mean value, $E(T)$, of the time to failure T. Hence,

$$\text{MTTF} = \int_0^\infty t \cdot f(t)\, dt \tag{6}$$

The MTTF may be written directly in terms of the reliability by substituting Eq. 4 into 6 and integrating by parts,

$$\text{MTTF} = \int_0^\infty R(t)\, dt \tag{7}$$

2.2 Hazard Function

The hazard function, or hazard rate $h(t)$, and often denoted by failure rate $\lambda(t)$, measures the rate of change of the probability that a surviving component will fail in the next infinitesimal time interval Δt. It can be written as:

$$h(t) = \lim_{\Delta t \to 0} \frac{P\left[(t < T \le t + \Delta t) \mid T > t)\right]}{\Delta t} = \frac{f(t)}{R(t)} \tag{8}$$

Replacing $f(t)$ given by Eq. 4 into 8 leads to:

$$h(t) = \frac{1}{R(t)}\left[-\frac{d\,R(t)}{dt}\right] \tag{9}$$

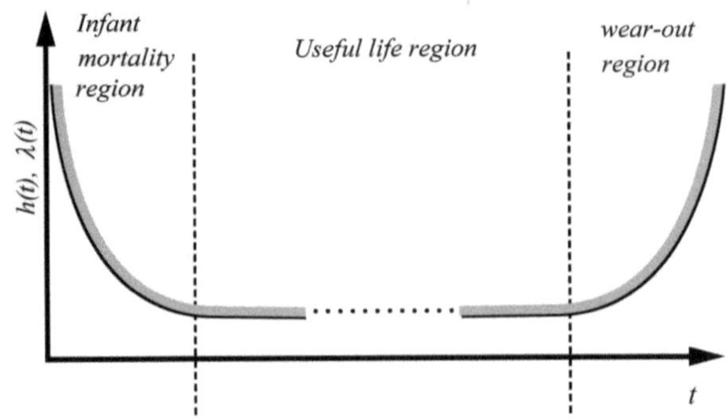

Fig. 2 Typical bathtub curve

from which after some manipulation it is possible to express the reliability in terms of the hazard function by:

$$R(t) = \exp\left[-\int_0^t h(t')\,dt'\right] \tag{10}$$

The failure density function $f(t)$ can also be obtained in terms of the hazard function simply by inserting Eq. 10 into 8,

$$f(t) = h(t)\exp\left[-\int_0^t h(t')\,dt'\right] \tag{11}$$

2.3 The Bathtub Curve

The hazard function of non-repairable components and systems follows a curve with bathtub shape, characterized by three distinct regions, as illustrated in Fig. 2. The period of time on the left-hand side of Fig. 2 is a region of high but decreasing failure rates. This is referred to as the period of *infant mortality*, or *early failures*. Most of the failures in the infant mortality region are quality related and result from inherent defects due to poor design, manufacturing and assembly. A substantial proportion of failures can also be attributed to a human error during installation and operation. Since most substandard components fail or their defects are detected and repaired and the experience of the personnel operating the equipment increases, the initially high failure rate gradually decreases. The preferred method for eliminating such failures is through design and production quality control measures that will reduce variability and hence susceptibility to infant mortality failures. If such measures are inadequate, a period of time may be specified during which the device undergoes wearin. During this time loading and use are controlled in such a way that weaknesses are likely to be detected and repaired without failure.

The middle section of the bathtub curve contains the smallest and most nearly constant failure rates and is referred to as the *useful life*. This flat behavior is characteristic of failures caused by random events and hence referred to as random failures. During this period of time, failures do not follow a predictable pattern and occur randomly due to the unexpected changes in stresses rather than from any inherent defect in the component or system under consideration. Consequently, the probability that a failure will occur in the next time increment is independent of the system's age.

In the *wear-out region*, the hazard rate increases with time as a result of irreversible aging effects. The failures are attributed to degradation or wear-out, which accumulates and accelerates over time. To minimize the failure effects, preventive maintenance or scheduled replacement of products is often necessary. Many components do not illustrate a complete bathtub curve. Instead, they have one or two segments of the curve. For example, most mechanical parts are dominated by the wear-out mechanism and thus have an increasing hazard rate. Some components exhibit a decreasing hazard rate in the early period, followed by an increasing hazard rate.

Each region can be modeled with a different reliability function. The main reliability model is the Weibull distribution that has the exponential distribution as one of the special cases. The Weibull functions can model a changing failure rate in time while the exponential distribution is used to model a constant failure rate in time (e.g., the useful life region of the bathtub curve). The Weibull is popular for modeling infant mortality and wear-out. Although the Weibull distribution is widely used in reliability [5] other distributions have also been used but are not discussed here due to space limitations. In general, the deciding factor in choosing a distribution type is to select the distribution function that best fits the data. In the following section, these reliability models are briefly introduced. Further information on these and other distributions may be found on any of the numerous references currently available on probability theory and statistical inference and on reliability engineering (e. g. [1, 4, 6]).

3 Common Reliability Models

3.1 Weibull Distribution

The Weibull distribution is one of the most widely used in reliability analysis. With an appropriate choice of its parameters all three regions of the bathtub curve can be represented. The three-parameter Weibull distribution is given by the distribution function:

$$F(t) = 1 - \exp\left[-\left(\frac{t-\tau}{\alpha}\right)^{\beta}\right] \qquad t \geq \tau \tag{12}$$

The parameters of the distribution are given by the set $\{\alpha, \beta, \tau\}$ with $\alpha, \beta > 0$ and $\tau \geq 0$. The parameters α, β and τ are the scale (or the characteristic life of the component), shape, and location parameters of the distribution, respectively. The two-parameter Weibull distribution (also referred to as *standard Weibull* model) is a special case of Eq. 12 with $\tau = 0$ so that:

$$F(t) = 1 - \exp\left[-\left(\frac{t}{\alpha}\right)^{\beta}\right] \qquad t \geq 0 \tag{13}$$

The corresponding probability density of the time to failure is

$$f(t) = \frac{d}{dt}F(t) = \frac{\beta}{\alpha}\left(\frac{t}{\alpha}\right)^{\beta-1} \exp\left[-\left(\frac{t}{\alpha}\right)^{\beta}\right] \quad t \geq 0 \tag{14}$$

The reliability function is

$$R(t) = \exp\left[-\left(\frac{t}{\alpha}\right)^{\beta}\right] \qquad t \geq 0 \tag{15}$$

and the hazard function is given by

$$h(t) = \frac{f(t)}{R(t)} = \frac{\beta}{\alpha}\left(\frac{t}{\alpha}\right)^{\beta-1} \tag{16}$$

The mean and variance of the time to failure are given by:

$$\mathrm{MTTF} = E(T) = \int_0^{\infty} R(t)\, dt = \alpha\,\Gamma\left(1 + \frac{1}{\beta}\right) \tag{17}$$

$$\mathrm{VAR}(T) = \alpha^2 \left[\Gamma\left(1 + \frac{2}{\beta}\right) - \Gamma^2\left(1 + \frac{1}{\beta}\right)\right] \tag{18}$$

where $\Gamma(\cdot)$ is the complete Gamma function defined by $\Gamma(x) = \int_0^{\infty} y^{x-1} e^{-y} dy$.

Figure 3 illustrates the Weibull probability density $f(t)$ and hazard $h(t)$ functions for $\alpha = 1$ and $\beta = 0.5$, 1, 1.5, and 2. As shown in the figure, the parameter β determines the shape of the distribution.

When $\beta < 1$, the Weibull distribution has a decreasing failure rate, which is typical of wearin phenomena, whereas for $\beta > 1$ the failure rate increases due to ageing effects of the components. When $\beta = 1$, the Weibull distribution has a constant failure rate and is reduced to the exponential distribution. When $\beta = 2$, the failure rate increases linearly with *t*, as shown in the $h(t)$ plot in Fig. 3. In this case the Weibull distribution is called the Rayleigh distribution.

The linearly increasing failure rate models are often used to model the life of some mechanical and electromechanical components, such as valves and electromagnetic relays, whose failure is dominated by mechanical or electrical wear-out. It can be seen that the Weibull distribution is very flexible and capable of

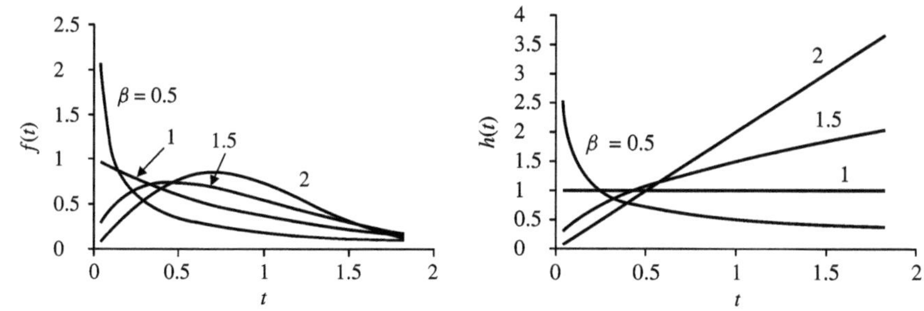

Fig. 3 Weibull $f(t)$ and $h(t)$ for $\alpha = 1$

modeling each region of a bathtub curve. It is the great flexibility that makes the Weibull distribution widely applicable.

3.2 Exponential Distribution

The exponential distribution is still the most widely used in reliability problems. As already referred, the exponential distribution is a particular case of a Weibull distribution with the shape parameter equal to one. The popularity of this distribution can be attributed primarily to the fact that it can be used to model the time to failure of components and systems with constant failure rate, a situation that is often realistic.

Replacing the hazard function $h(t)$ by a constant failure rate λ in Eq. 10, the reliability may be written as:

$$R(t) = \exp\left[-\lambda t\right] \tag{19}$$

Similarly, the CDF and the PDF, become:

$$F(t) = P[T \leq t] = 1 - \exp\left[-\lambda t\right] \tag{20}$$

$$f(t) = \lambda \exp\left[-\lambda t\right] \tag{21}$$

Equation 20 is denoted as the "exponential distribution". Its MTTF is the mean of the distribution, given by:

$$\mathrm{MTTF} = \int_0^\infty t \cdot f(t)\, dt = \frac{1}{\lambda} \tag{22}$$

If the time to failure of a component is exponentially distributed, the probability of failure in a specified time interval does not depend on the age of the component. The probability that the component will fail within a specified time interval is the same, irrespective of whether the component has been used for some time or has just been placed in use. In other words, the probability that the life will be greater than $t + \Delta t$, given that the component has survived time t does not depend on t. The component is as-good as-new. This characteristic of the exponential model is called the memoryless property. Mathematically, this property can be expressed as:

$$P[T > t + \Delta t | T > t] = \frac{R(t + \Delta t)}{R(t)} = \frac{\exp[-\lambda(t + \Delta t)]}{\exp[-\lambda t]} = \exp[-\lambda \Delta t] \tag{23}$$

The memoryless property of the exponential distribution makes this model suitable for components whose conditional probability of failure within a specified time interval practically does not depend on age, i.e. for components which practically do not degrade or wear-out with time.

3.3 Models Involving Two or More Distributions

In many situations the equipments being studied are prone to different failure modes and thus the failure data cannot be adequately modeled by a single distribution. Then, a natural alternative is to examine models involving two or more distributions. Several models involving two or more Weibull distributions have been developed (e.g. [5]). These include mixture, competing risk, multiplicative and composite models.

Weibull Mixture Model

In general, a mixed distribution is defined by a combination of two or more distributions,

$$F(t) = \sum_{i=1}^{n} p_i F_i(t) \quad with \quad p_i > 0 \text{ and } \sum_{i=1}^{n} p_i = 1 \tag{24}$$

where the distributions involved are called the *subpopulations*.

In the case the $F_i(t)$ s, are either two or three-parameter Weibull distributions, the model is called *finite Weibull mixture model*. In the literature, the Weibull mixture model is also referred to by many other names, such as, additive-mixed Weibull distribution, bimodal-mixed Weibull (for a two fold mixture), mixed-mode Weibull distribution, Weibull distribution of the mixed type, multimodal Weibull distribution, etc.

Mixture arises when the population of interest contains two or more nonhomogeneous subpopulations, which occurs frequently in practice. Examples of such cases occur when a subpopulation is mixed with a substandard subpopulation, due to manufacturing process variation and material defects, which will fail in early time. Another example that usually produces nonhomogeneous subpopulations is associated with the use of components from different suppliers.

A special case of interest is the mixture comprising two Weibull distributions given by Eq. 13:

$$F(t) = pF_1(t) + (1-p)F_2(t) \tag{25}$$

In this case the model is characterized by five parameters: the shape and scale parameters for the two subpopulations $\{\alpha_1, \beta_1, \alpha_2, \beta_2\}$ and the mixing parameter $p\,(0<p<1)$. The density and the hazard functions are given by:

$$f(t) = pf_1(t) + (1-p)f_2(t) \tag{26}$$

and

$$h(t) = \frac{pR_1(t)}{pR_1(t) + (1-p)R_2(t)}h_1(t) + \frac{(1-p)R_2(t)}{pR_1(t) + (1-p)R_2(t)}h_2(t) \tag{27}$$

Competing Risk Model

The competing risk model evaluates the component reliability by "building up" from the reliability models for each failure mode. The model assumes that each failure mechanism leading to a particular type of failure (i.e., failure mode) proceeds independently of every other one, at least until a failure occurs. Moreover, the component fails when the first of all the competing failure mechanisms reaches a failure state.

A general n-fold competing risk model involving n subpopulations is given by:

$$F(t) = 1 - \prod_{i=1}^{n}[1 - F_i(t)] \tag{28}$$

In the reliability literature, the competing risk model is also called the compound model, series system model, multirisk model, and poly-Weibull model (if the subpopulations are Weibull distributions).

Multiplicative models

A general n-fold multiplicative model, also referred to as *complementary risk model*, is given by:

$$F(t) = \prod_{i=1}^{n} F_i(t) \tag{29}$$

The model can be used in the context of modeling a functionally parallel system with independent components, but in contrast to the competing risk model, this model has received very little attention in the literature.

Composite Models

In a composite model (also known as sectional model, piecewise model, and step function model), the failure distributions over different time intervals are given by different distribution functions. As a result, $F(t)$ over n successive intervals is given by:

$$F(t) = \begin{cases} k_1 F_1(t) & \text{for } t \in (0,\ t_1) \\ & \cdots\cdots \\ k_{n-1} F_{n-1}(t) & \text{for } t \in (t_{n-2},\ t_{n-1}) \\ (1 - k_n) + k_n F_n(t) & \text{for } t \in (t_{n-1},\ \infty) \end{cases} \tag{30}$$

where $F_i(t)$ s are the distribution functions of each of the n subpopulations. The time instants t_i, $1 \le i \le (n-1)$, are referred to as the break (partition) points, and they form an increasing sequence. The k_i s are all > 0 and they define a family of models. For the distribution and density functions to be continuous at the break points, the parameters need to be constrained. This model offers greater flexibility and is suited to modeling a variety of complex failure data.

In the case of two subpopulations $(n = 2)$, $k_1 = k_2 = 1$ and if $F_1(t)$ is a two-parameter Weibull distribution and $F_2(t)$ a three-parameter Weibull distribution with location parameter τ, the continuity conditions at t_1 require:

$$F_1(t_1^-) = F_2(t_1^+) \quad \text{and} \quad f_1(t_1^-) = f_2(t_1^-) \tag{31}$$

which result in

$$t_1 = \left[\frac{\alpha_1^{\beta_1}}{\alpha_2^{\beta_2}} \left(\frac{\beta_2}{\beta_1} \right)^{\beta_2} \right]^{1/(\beta_1 - \beta_2)} \quad \text{and} \quad \tau = \left(1 - \frac{\beta_2}{\beta_1} \right) t_1 \tag{32}$$

The above conditions indicate that although the composite model has six parameters, only four are independent.

4 System Reliability Modeling and Analysis

From the hierarchical structure point of view, a system is comprised of a number of subsystems, which may be further divided into lower-level subsystems, depending on the purpose of system analysis. Basic components are the lowest-level constituents of a system. This chapter describes the methods for evaluating the reliability of systems with different configurations.

4.1 Fault Tree Analysis (FTA)

The fault tree technique was introduced in 1962 at Bell Telephone .Laboratories, in connection with a safety evaluation of the launching system for the intercontinental Minuteman missile. The Boeing Company improved the technique and introduced computer programs for both qualitative and quantitative fault tree analysis. Today fault tree analysis is one of the most commonly used techniques for risk and reliability studies.

FTA is a top-down process by which an undesirable event, referred to as the top event (e.g., a failure of the system to accomplish its function), is logically decomposed into possible causes in increasing detail to determine the causes or combinations of causes of the top event. A basic event is defined as an event in a fault tree model that requires no further development, because the appropriate limit of resolution has been reached.

A FTA can yield both qualitative and quantitative information about the system under study. Qualitative information may include failure paths, root causes, and weak areas of the system. Quantitative analysis of a fault tree gives a probabilistic estimation of the top event and can lead to a conclusion as to whether the design is adequate in terms of reliability and safety.

The first step of a fault tree analysis clearly consists in the definition of the top event to be analyzed and of the boundary conditions for the analysis. Then fault events that are the immediate, necessary, and sufficient causes that result in the top event are identified and connected to the top event via logic gates. Logic symbols graphically represent the gates used to interconnect the low-level events that contribute to the top-level event according to their causal relations. Figure 4 lists the most commonly used fault tree logic symbols together with a brief description of their interpretation.

Boolean and Structure Functions

A fault tree can be considered a graphical representation of Boolean relationships among fault tree events that cause the top event to occur. Therefore, it is possible to translate a fault tree into an entirely equivalent Boolean function by using the rules of Boolean algebra shown in Fig. 5. The following operators and rules of Boolean algebra are commonly used in FTA. More information on Boolean rules and functions of fault trees may be found in (e.g. [3, 7]).

However, in quantitative fault tree analysis it is convenient to replace conjunction, disjunction, and negation by (arithmetical) addition, subtraction, and multiplication leading to the so-called *structure function of the system* or just the

Name	Description	Logic Symbols
AND gate	Output event occurs if all input events occur simultaneously.	
OR gate	Output event occurs if any one of the input events occurs.	
INHIBIT gate	Input produces output when a conditional event occurs.	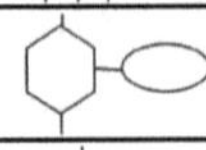
EXCLUSIVE OR gate	Output event occurs if only one of the input events occurs.	
VOTING gate	Output event occurs if at least *k* of *n* input events occur.	

Fig. 4 Fault tree logical symbols

Fig. 5 Boolean operators and rules

Boolean Operators	
∧ Conjunction (AND operation)	
∨ Disjunction (OR operation)	
− Negation	
Boolean Rules	
Commutative	X ∧ Y = Y ∧ X X ∨ Y = Y ∨ X
Idempotent	X ∧ X = X X ∨ X = X
Associative	X ∧ (Y ∧ Z) = (X ∧ Y) ∧ Z X ∨ (Y ∨ Z) = (X ∨ Y) ∨ Z
Absorption	X ∧ (X ∨ Y) = X X ∨ (X ∧ Y) = X
Identity	1 ∧ X = X 0 ∨ X = X
Distributive	X ∧ (Y ∨ Z) = X ∧ Y ∨ X ∧ Z X ∨ (Y ∧ Z) = (X ∨ Y) ∧ (X ∨ Z)

structure. The structure function ψ of the fault tree is an algebraic binary function that returns the value in the following way:

$$\psi(\mathrm{X}) = \begin{cases} 1 & \text{if top event occurs (failure of the system)} \\ 0 & \text{otherwise} \end{cases} \tag{33}$$

The state of component *i*, for $i = 1, 2, \ldots, n$ is described by a binary variable X_i, where:

$$X_i = \begin{cases} 1 & \text{if basic event } i \text{ occurs (component fails)} \\ 0 & \text{otherwise} \end{cases} \tag{34}$$

The top event probability P_{fs}, referred to as *probability of failure* or *unreliability* of the system, is then the probability of the structure function taking the value of unity, which is calculated by an expected value of the structure function:

$$P_{f_s} = P\,[\text{top event}] = P\,[\psi(\mathrm{X}) = 1] = E\,[\psi(\mathrm{X})] \tag{35}$$

If the basic event i means that component i is in a failed state, then

$$P[X_i = 1] = E[X_i] = P_{f_i} = 1 - R_i \tag{36}$$

where P_{f_i} is the unreliability of component i and R_i is the probability that component i is functioning.

In a AND gate, the top event occurs (system fails) if and only if all the basic events occur (components fail) simultaneously. Then, the structure function of the AND gate can be expressed as the algebraic form:

$$\psi(\mathrm{X}) = X_1 \cdot X_2 \ldots X_n = \prod_{i=1}^{n} X_i \tag{37}$$

Assuming that the basic events are independent, the corresponding probability is calculated from:

$$\begin{aligned} P_{f_s} &= P\,[\psi(\mathrm{X}) = 1] = E\,[\psi(\mathrm{X})] = E\,[X_1 \cdot X_2 \ldots X_n] \\ &= E\,[X_1]E\,[X_2]\ldots E\,[X_n] = \prod_{i=1}^{n} p_{f_i} \end{aligned} \tag{38}$$

In a OR *gate*, the top event occurs if at least one of the basic events occurs. Thus, the structure function of the OR gate can be expressed as the algebraic form:

$$\psi(\mathrm{X}) = 1 - (1 - X_1) \cdot (1 - X_2) \ldots (1 - X_n) = 1 - \prod_{i=1}^{n} (1 - X_i) \tag{39}$$

Assuming that the basic events are independent, the corresponding probability is calculated from:

$$\begin{aligned} P_{f_s} = P\,[\psi(\mathrm{X}) = 1] = E\,[\psi(\mathrm{X})] &= 1 - \prod_{i=1}^{n} E\,[1 - X_i] = 1 - \prod_{i=1}^{n} (1 - E[X_i]) \\ &= 1 - \prod_{i=1}^{n} (1 - p_{f_i}) \end{aligned} \tag{40}$$

After developing the structure function of the top event by combining Eqs. 37 and 39, the exact probability of system failure is then obtained by deleting the powers k of X_i s (i.e. $X_i^k = X_i$ from $i = 1$ to n) and by replacing X_i s by the corresponding probabilities p_{f_i}.

Structures Represented by Path and Cut Sets

When the complexity of the calculation of the top event probability by means of the structure function of the fault tree increases, approximations can be derived based on the minimal cut and path sets of the system.

A *cut set* (C_j) in a fault tree is a set of basic events (component failures) whose occurrence (at the same time) ensures that the top event occurs (failure of the system).

Some cut sets may contain unnecessary components. If removed, failure of the remaining components still results in a system failure. Such cut sets can be further reduced to form *minimal cut sets*. Therefore a minimal cut set is the smallest combination of components which if they all fail will cause the system to fail. If any component is removed from the set, the remaining components collectively are no longer a cut set.

A *path set* (P_j) is a set of basic events whose nonoccurrence (at the same time) ensures that the top event does not occur. A path set is said to be a *minimal path set* if the set cannot be reduced without losing its status as a path set.

Minimal Cut Set Representation

If the system has *nc* minimal cut sets, the top event can be represented by a OR gate with the minimal cut sets as input. Each minimal cut set occurs if all its basic events occur simultaneously and, therefore, the basic events in a minimal cut set are associated by a AND gate.

The minimal cut set representation of the structure function the top event is then:

$$\psi(\mathrm{X}) = 1 - \prod_{j=1}^{nc} (1 - C_j(\mathrm{X})) \tag{41}$$

where the structure function $C_j(\mathrm{X})$ of the minimal cut set j with n_j basic events is given by:

$$C_j(\mathrm{X}) = \prod_{i \in C_j}^{n_j} X_i \tag{42}$$

Minimal-Path Set Representation

Similarly, if the system has *np* minimal path sets, the top event can be represented by a AND gate with the minimal path sets as input. Each minimal path set *j* occurs if at least one of its basic events occur and, therefore, the basic events in a minimal path set are associated by a OR gate.

The minimal path set representation of the structure function the top event is then:

$$\psi(\mathrm{X}) = \prod_{j=1}^{np} P_j(\mathrm{X}) \tag{43}$$

where the structure function $P_j(\mathrm{X})$ of the minimal path set j with n_j basic events is given by:

$$P_j(\mathrm{X}) = 1 - \prod_{i \in P_j}^{n_j} (1 - X_i) \tag{44}$$

Upper and Lower Bounds of System Probability of Failure

Although the basic events are assumed to be independent, the same basic event may occur in several minimal cut sets. In these cases the product operation in the minimal cut representation cannot be interchanged by the expected-value operation because terms $(1 - C_j(\mathrm{X}))$ for different j are statistically dependent. Therefore the minimal cut set representation of the structure function only provides an *upper bound approximation* $(P_{f_s}^{ub})$ (conservative) for the probability of system failure (P_{f_s}). Similarly, a *lower bound approximation* (unconservative) for the probability of system failure $(P_{f_s}^{lb})$ is obtained when product operation in a minimal path set representation is interchanged by the expected-value operation,

$$P_{f_s}^{lb} = \prod_{j=1}^{np} E[P_j(\mathrm{X})] \quad \leq \quad P_{f_s} = E[\psi(\mathrm{X})] \quad \leq P_{f_s}^{ub} = 1 - \prod_{j=1}^{nc} (1 - E[C_j(\mathrm{X})]) \tag{45}$$

The equality holds when each basic event appears in exactly one minimal path or cut set. When one or more components appear in more than one minimal cut or path sets, the exact probability of system failure can be obtained by repeated pivotal decomposition of the structure function with respect to each variable X_i (and deleting the powers k of X_i s (i.e. $X_i^k = X_i$ from i = 1 to n) by:

$$\psi(\mathrm{X}) = X_i \psi(1_i, \mathrm{X}) + (1 - X_i)\, \psi(0_i, \mathrm{X}) \tag{46}$$

4.2 Reliability Block Diagram

A reliability block diagram (RBD) is a success-oriented network describing the *function* of the system. It shows the logical connections of (functioning) components needed to fulfill a specified system function. In a reliability block diagram, components are symbolized by rectangular blocks, which are connected by straight lines according to their logic relationship. Depending on the purpose of the system analysis, a block may represent a lowest-level component, a module, or a subsystem.

In a reliability block diagram, connection through a block means that the component represented by the block is functioning. If the system is constituted by n components it is said that the system works (i.e. the specified system function is achieved) if the connection is established among the end points of the reliability block diagram.

In some practical applications, the model of the system can be constructed alternatively by a fault tree or by a reliability block diagram. When the fault tree is limited to only OR- and AND-gates, both methods may yield the same result, and it is possible to convert the fault tree to a reliability block diagram, and vice versa. An AND gate in a fault tree is logically equivalent to a parallel reliability block diagram, both describing the same logic that the top event occurs only when all contributing causes occur. An OR gate in a fault tree logically corresponds to a series reliability block diagram because both graphically represent the same logic that the top event occurs when one of the basic events occurs. The relationship between some simple reliability block diagrams and fault trees is illustrated in Fig. 6.

The conversion of a fault tree to a reliability block diagram usually starts from the bottom of the tree. The basic events under the same gate at the lowest-level of the tree form a block depending on the type of the gate. The block is treated as a single event under the next high-level gate. The block, along with other basic events, generates an expanded block. This expanded block is again considered as a single event, and conversion continues until an intermediate event under a gate is seen. Then the intermediate event is converted to a block by the same process. The block and existing blocks, as well as the basic events, are put together according to the type of the gate. The process is repeated until the top gate is converted.

4.3 Reliability of Series Systems

A reliability block diagram is in a series configuration when failure of any one block (according to the failure mode of each item based on which the reliability block diagram is developed) results in the failure of the system. Accordingly, for functional success of a series system, all of its blocks (items) must successfully function during the intended mission time of the system. The reliability of the system is the probability that all blocks succeed during its intended mission time t. Thus, probabilistically, the system reliability for independent blocks is obtained from:

$$R_s(t) = R_1(t) \cdot R_2(t) \cdot R_3(t) \cdots R_n(t) = \prod_{i=1}^{n} R_i(t) \tag{47}$$

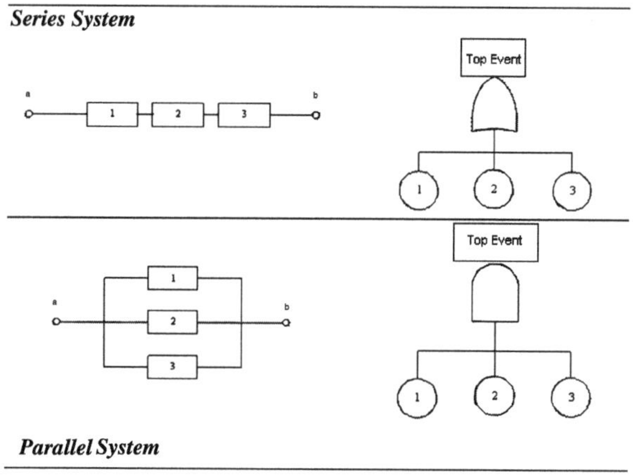

Fig. 6 Relationship between some simple reliability block diagrams and fault trees

which indicates that the system reliability is the product of components reliabilities $R_i(t)$. Since $R_i(t) \leq 1$, this results in the system reliability being less than the reliability of any component. Furthermore, the system reliability decreases rapidly as the number of components in a system increases.

Considering the simple case where the times to failure of n components in a system are modeled with the exponential distribution, where λ_i is the failure rate of component i, then from Eq. 47, the system reliability can be written as

$$R_s(t) = \exp\left(-t\sum_{i=1}^{n}\lambda_i\right) = \exp(-\lambda_s t) \tag{48}$$

where λ_s is the failure rate of the system, $\lambda_s = \sum_{i=1}^{n} \lambda_i$.

The mean time to failure of the series system is

$$\text{MTTF} = \int_0^\infty R(t)\,dt = \frac{1}{\sum_{i=1}^{n}\lambda_i} \tag{49}$$

4.4 Reliability of Active Parallel Systems

A system is said to be an *active parallel system* if and only if the failure of all components within the system results in the failure of the entire system. In other words, a parallel system succeeds if one or more components are operational. In contrast to a series system, the reliability of an active parallel system increases with the number of components within the system. Thus, a parallel configuration is a method of increasing system reliability. For a general active parallel configuration of n independent components, the unreliability (probability of failure) of the system is calculated by:

$$F_a(t) = F_1(t) \cdot F_2(t) \cdot F_3(t) \cdots F_n(t) \tag{50}$$

Since $R_i(t) = 1 - F_i(t)$, then:

$$R_a(t) = 1 - F_a(t) = 1 - \prod_{i=1}^{n} [1 - R_i(t)] \tag{51}$$

When the system is composed of only two components with constant failure rate (exponential time to failure model), the reliability is given by:

$$R_a(t) = \exp(-\lambda_1 t) + \exp(-\lambda_2 t) - \exp[-(\lambda_1 + \lambda_2)\, t] \tag{52}$$

and the MTTF is

$$\mathrm{MTTF} = \int_0^\infty R_a(t)\, dt = \frac{1}{\lambda_1} + \frac{1}{\lambda_2} - \frac{1}{\lambda_1 + \lambda_2} \tag{53}$$

In the special case of n identical components with a constant failure rate λ (and $\mathrm{MTTF} = 1/\lambda$) in an active parallel configuration, Eq. 51 simplifies to the following form:

$$R_a(t) = 1 - [1 - \exp(-\lambda t)]^n \tag{54}$$

and the $\mathrm{MTTF_a}$ of the system is:

$$\mathrm{MTTF_a} = \mathrm{MTTF} \sum_{i=1}^{n} \frac{1}{i} \tag{55}$$

It can be seen that in the design of active parallel systems, the MTTF of the system exceeds the MTTF of the individual components. However, the contribution to the MTTF of the system from the second unit, the third unit, and so on would have a diminishing return as n increases. That is, there would be an optimum number of parallel blocks (units) by which a designer can maximize the reliability and at the same time minimize the cost of the component in its life cycle.

In practice, the benefit in terms of reliability of the active parallel systems can significantly be reduced due to phenomena that create dependencies between the failure of two or more redundant components. This limitation of the active parallel systems is typically attributed to *common-mode failures*, which cause them to fail simultaneously. When the time to failure of the components follows an exponential distribution, such phenomena are modeled by assuming that a fraction β of the total failure rate of a single component λ is attributed to common-mode failures:

$$\lambda = \lambda_c + \lambda_I = \beta\lambda + (\beta - 1)\lambda \tag{56}$$

where λ_I is the rate of independent failure and λ_c is the common-mode failure rate.

The reliability of the system is then given by the reliability of the active parallel configuration of components with independent failures (calculated with λ_I) multiplied by $\exp(-\lambda_c t)$ to account for common-mode failures. Thus, for the two units in parallel, the reliability is:

$$\begin{aligned} R_a'(t) &= \exp(-\lambda_c t) \cdot [2\exp(-\lambda_I t) - \exp(-2\lambda_I t)] \\ &= [2 - \exp(-(1-\beta)\lambda t)]\ \exp(-\lambda t) \end{aligned} \tag{57}$$

The effect of common-mode failures can also be seen in the reduction in the mean time to failure as a function of the parameter β, given by:

$$\mathrm{MTTF}_{\mathrm{a}}' = \left[2 - \frac{1}{2-\beta}\right] \mathrm{MTTF} \tag{58}$$

4.5 k-out-of-n Systems

Some systems require more than one component to work in order for the entire system to operate. Such systems are usually referred to as *k-out-of-n systems*, where n is the total number of components in the system and k is the minimum number of components that must function for the system to operate successfully. Assuming that all units are independently and identical with reliability $R(t)$, the number of operational components in the system (x), is a random variable that follows a binomial distribution. Therefore, the reliability of the k-out-of-n system corresponds to the probability of having k components operational,

$$R_{k/n}(t) = P(x \geq k) = \sum_{i=k}^{n} C_n^i R(t)^i (1 - R(t))^{n-i} \tag{59}$$

If the time to failure of the components follows an exponential distribution, the system reliability is:

$$R_{k/n}(t) = \sum_{i=k}^{n} C_n^i \exp(-\lambda i t)(1 - \exp(-\lambda t))^{n-i} \tag{60}$$

where λ is the component failure rate. The mean time to failure of the system is

$$\mathrm{MTTF}_{\mathrm{a}} = \mathrm{MTTF} \sum_{i=k}^{n} \frac{1}{i} \tag{61}$$

4.6 Standby Systems

A system is called a *standby redundant system* when one or more standby components or subsystems enable the system to continue the function when the

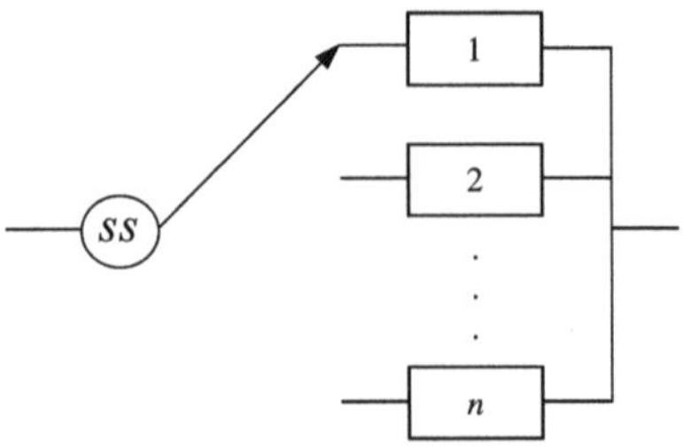

Fig. 7 Standby redundant system

primary unit fails. Figure 7 illustrates a standby system consisting of n components, where component 1 is the primary component and ss represents the switching system.

In this configuration, unit 1 operates permanently until it fails. The sensing and switching device ss recognizes a unit failure in the system and switches to another unit. This process continues until all standby units have failed, in which case the system is considered failed. Since units 2 to n do not operate constantly (as is the case in active parallel systems), they would fail at a much lower rate. This is because the failure rate for components is usually lower when the components are operating than when they are idle or dormant. It is clear that system reliability is totally dependent on the reliability of the sensing and switching device.

The reliability of a redundant standby system is the reliability of unit 1 over the mission time t (i.e., the probability that it succeeds the whole mission time) plus the probability that unit 1 fails at time t, prior to t and the probability that the sensing and switching unit does not fail by t, and the probability that standby unit 2 does not fail by t (in the standby mode) and the probability that standby unit 2 successfully functions for the remainder of the mission in an active operation mode, and so on. Mathematically, the reliability function for a two block (unit) standby perfect switching system according to this definition can be obtained as:

$$R_{sb}(t) = R_1(t) + \int_0^t R_2(t - t')f_1(t')dt' \tag{62}$$

where $f_1(t')$ is the pdf of the time to failure of component 1 and $R_2(t - t')$ is the reliability of unit 2 after it started to operate at time t'.

In a more general case in which both, the reliability of the sensing and switching device $R_{ss}(t')$ and the reliability of unit 2 in the standby mode $R_2^{+}(t')$ are taken into account, the reliability of the system is given by:

$$R_{sb}(t) = R_1(t) + \int_0^t R_2(t - t')f_1(t')\, R_{ss}(t')R_2^{+}(t')dt' \tag{63}$$

When the time to failure of all components follows and exponential distribution the reliability of the standby system is given by:

$$R_{sb}(t) = \exp(\lambda_1 t) + \int_0^t [\exp(-\lambda_2(t-t'))]\,[\lambda_1 \exp(-\lambda_1 t')] \exp(-\lambda_{ss} t') \exp(\lambda_2^+ t')\, dt'$$

$$R_{sb}(t) = \frac{\lambda_1 \exp(-\lambda_2 t)}{\lambda_1 + \lambda_{ss} + \lambda_2^+ - \lambda_2} \{ 1 - \exp[-(\lambda_1 + \lambda_{ss} + \lambda_2^+ - \lambda_2)\, t] \} \tag{64}$$

For the special case of two identical components $\lambda_1 = \lambda_2 = \lambda$ and probability P of failure of the sensing and switching device, the reliability of the system is given by,

$$R_{sb}(t) = \left[1 + (1-p) \frac{\lambda}{\lambda_2^+} (1 - \exp(-\lambda_2^+ t)) \right] e^{-\lambda t} \tag{65}$$

which can be further simplified for perfect sensing and switching, $p = 0$ (or $\lambda_{ss} = 0$), and no standby failures, $\lambda_2^+ = 0$,

$$R_{sb}(t) = (1 + \lambda t) \exp(-\lambda t) \tag{66}$$

In the design of standby parallel systems it is necessary to consider three types of configurations, namely *cold*, *warm*, or *hot* standby, which represent a trade-off between switching failures and failure in standby. In cold standby the secondary unit is shut down until needed. This typically reduces the value of λ_2^+ to a minimum. However, it tends to result in the largest values of p. However, coming from cold startup to a fully loaded operation on short notice may cause sufficient transient stress to result in a significant demand failure probability. In warm standby the transient stresses are reduced by having the secondary unit continuously in operation, but in an idling or unloaded state. In this case p is reduced and λ_2^+ moderately increases. The switching failure probability p can be further reduced by having the secondary unit in hot standby, that is, continuously operating at a full load. For this situation (and in this case of identical units) the failure rate λ^+ will be equal that of the primary unit λ.

4.7 Reliability Evaluation of Complex Systems

Reduction Method

Real systems are frequently a combination of independent parallel and series configurations denoted as *parallel-series systems*. A parallel-series system can be analyzed by dividing it into its basic parallel and series modules and then

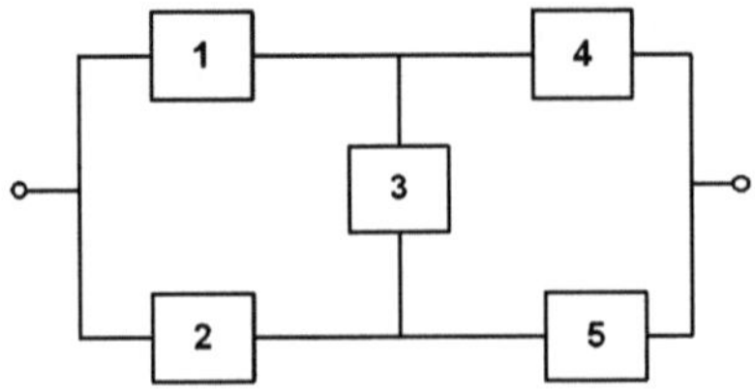

Fig. 8 Nonparallel-series bridge system

determining the reliability function for each module separately using Eqs. 51 and 47, respectively. The process continues until the system is reduced to one single block with reliability equivalent to the whole system. This method of analyzing parallel-series systems is known as *reduction method*.

Decomposition Method

In some situations the linkage of the components or subsystems is such that the foregoing technique of decomposing into parallel and series configurations can not be applied directly. Such is the case for the bridge system configuration shown in Fig. 8.

This and similar systems can be analyzed by decomposing the system into a combination of series-parallels by using the total probability rule. In this approach, known as *decomposition method*, the reliability of a system is equal to the reliability of the system given that a chosen unit (e.g., unit 3 in Fig. 8) is working times the reliability of the unit (3), plus the reliability of the system given that the chosen unit (3) is failed times the unreliability of the unit (3), i.e.:

$$R_s(t) = R_s(t|3\,works)\,R_3(t) \;+\; R_s(t|3\,fails)\,(1 - R_3(t)) \tag{67}$$

In general, Eq. 67 is applied to all units that make the system a nonparallel-series system so that each of the conditional reliability terms in Eq. 67 represents a purely parallel-series system for which the reliability determination is simple.

Minimal Cut Sets Method

An alternative approach for determining the reliability of a complex system involves the identification of the minimal cut sets of the system. Since the system works if none of the minimal cut sets occur, assuming that the nc minimal cut sets C_i of the system are independent, a lower bound of the reliability of the system can be calculated from:

$$R_s^{lb} = \prod_{i=1}^{nc} (1 - P(C_i)) \tag{68}$$

where C_i ($i = 1, 2, \ldots, nc$) represents the event that components in minimal cut set i are all in a failure state and nc is the total number of minimal cut sets.

As already referred, Eq. 68 only provides a lower bound (conservative estimation) of the system reliability since in most of the cases the same component can appear in several minimal cut sets and, therefore, the occurrence of the minimal cut sets is not independent, as assumed in Eq. 68. Thus, the product in Eq. 68 is an underestimation of the probability of no cut set failures.

The exact value of the reliability of the system can be obtained by using the inclusion–exclusion rule and the rules of Boolean algebra. Since every minimal cut set causes the system to fail, the event that the system fails is the union of all minimal cut sets. Then the system reliability can be written as:

$$R_s = 1 - P(C_1 \cup C_2 \cup C_3 \cup \ldots\ldots \cup C_{nc}) \tag{69}$$

Equation 69 is then evaluated by applying the inclusion–exclusion rule, which is given by:

$$\begin{aligned} P(C_1 \cup C_2 \cup C_3 \cup \ldots \cup C_{nc}) = & \sum_{i=1}^{nc} P(C_i) - \sum_{i<j=2}^{nc} P(C_i \cap C_j) + \\ & + \sum_{i<j<k=3}^{nc} P(C_i \cap C_j \cap C_k) + \ldots \\ & + (-1)^{nc-1} P(C_i \cap C_j \ldots \cap C_{nc}) \end{aligned} \tag{70}$$

As an example, consider a 2-out-of-3 system of identical components with probability of failure $P_f = 0.2$. Minimal cut sets of the system are $\{1,2\}$, $\{1,3\}$ and $\{2,3\}$. Let A_i denote the event that component i has failed, $i = 1, 2, 3$. Then the events described by the minimal cut sets can be written as $C_1 = A_1 \cap A_2$, $C_2 = A_1 \cap A_3$ and $C_3 = A_2 \cap A_3$. From Eqs. 69 and 70, and using the rules of Boolean algebra (namely the identity law $A \cap A = A$) the system reliability can be evaluated by:

$$\begin{aligned} R_s &= 1 - P(C_1 \cup C_2 \cup C_3) = 1 - [P(C_1) + P(C_2) + P(C_3) \\ &\quad - P(C_1 \cap C_2) - P(C_1 \cap C_3) - P(C_2 \cap C_3) + P(C_1 \cap C_2 \cap C_3)] \\ &= 1 - [P(A_1 \cap A_2) + P(A_1 \cap A_3) + P(A_2 \cap A_3) \\ &\quad - P(A_1 \cap A_2 \cap A_3) - P(A_1 \cap A_2 \cap A_3) - P(A_1 \cap A_2 \cap A_3) + P(A_1 \cap A_2 \cap A_3)] \\ &= 1 - \left[3(P_f^2) - 3(P_f^3) + (P_f^3)\right] = 0.896 \end{aligned} \tag{71}$$

which is larger than the lower bound for the reliability of the system provided by Eq. 68, $R_s^{lb} = \left(1 - P_f^2\right)^3 = 0.885$.

5 Repairable Systems

Nonrepairable items are those that are discarded and replaced with new ones when they fail. Reliability of a nonrepairable item is expressed in terms of its time-to-failure distribution, which can be represented by respective cdf, pdf, or hazard (failure) rate function, as was discussed previously. However, relatively few systems are designed to operate without maintenance of any kind. Maintenance consists of all activities performed on an item to retain it in or to restore it to a specified state. Usually one distinguishes two types of maintenance: *preventive maintenance*, and *corrective maintenance* also known as repair.

Preventive maintenance (*PM*) is planned maintenance performed when an item is functioning properly to prevent future failures. It may involve inspection, adjustments, lubrication, parts replacement, calibration, and repair of items that are beginning to wear-out. PM is generally performed on a regular basis, regardless of whether or not functionality or performance is degraded. The aim of preventive maintenance must also be to detect and repair hidden failures, i.e. failures in redundant elements.

Corrective maintenance (*CM*) often called repair is carried out after an item has failed. The purpose of corrective maintenance is to bring the item back to a functioning state as soon as possible, either by repairing or replacing the failed item or by switching in a redundant item.

Typically models of preventive maintenance consider both the case of ideal maintenance, in which the system is restored to an as-good-as-new condition each time maintenance is performed, and also more realistic situations of imperfect maintenance that includes the effect of human reliability on the overall reliability of a maintained system. Although models are available for these maintenance actions, in the following section only two simple corrective maintenance models are presented: replacement after failure and periodic testing/replacement.

5.1 Availability and Maintainability of Repairable Systems

When studying repairable systems the probability of failure is no longer the most important characteristic of interest. For those systems the number of failures and the time required to make repairs are also important and two new reliability parameters become the focus of attention: the *Availability* and *Maintainability*.

The notion of *availability*, is related to repairable (or maintained) items only and is defined as the probability that a repairable system (or component) will function at time t. Respectively, the unavailability of a repairable system (or component) is defined as the probability that the item is in a failed state (down) at time t, which depends on the number of failures and repair times. *Maintainability* is a measure of how fast a system may be repaired following failure.

There are several definitions of *availability*, the most common ones are as follows:

1. Instantaneous (point) availability of a repairable item at time t, $A(t)$, is the probability that the system (or component) is up at time t;
2. Average or interval availability, $\bar{A}(T)$ defined for a fixed time interval, T, as

$$\bar{A}(T) = \frac{1}{T} \int_0^T A(t)\, dt \qquad (72)$$

which is just the value of the point availability averaged over some interval of time T. This may be the design life of the system or the time to accomplish some particular mission;
3. The steady-state, asymptotic availability or limiting average availability, defined as:

$$A_\infty = \lim_{T \to \infty} \frac{1}{T} \int_0^T A(t)\, dt \qquad (73)$$

A non-repairable system is available until a failure occurs and, therefore, its availability coincides with its reliability function $R(t)$,

$$A(t) = R(t) \qquad (74)$$

Introducing this expression into the former equation and considering the limit $T \to \infty$, the MTTF (mean time to failure) is obtained from the numerator. On other hand, the denominator tends to infinite. Thus the steady-state availability of a non-repairable system is $A_\infty = 0$.

Maintainability is a characteristic of an item, expressed by the probability that a preventive maintenance or a repair of the item will be performed within a stated time interval for given procedures and resources (skill level of personnel, spare parts, test facilities, etc.). If T_r is a random variable that characterizes the (random) time required to carry out a repair, then the repair time distribution function or simply, the maintainability, $M(t)$, is:

$$M(t) = P[T_r \leq t] \qquad (75)$$

and the mean time to repair (MTTR) is:

$$\text{MTTR} = \int_0^\infty t\, m(t)\, dt \qquad (76)$$

where $m(t) = dM(t)/dt$ is the repair time density probability function.

The repair rate $\mu(t)$ is obtained from the probability that the component is repaired between t and $t + \Delta t$, given that it failed in time instant t:

$$\mu(t)\,\Delta t = \frac{P[t \le T_r \le t + \Delta t]}{P[T_r > t]} \tag{77}$$

Since $m(t)\Delta t = P[t \le T_r \le t + \Delta t]$, the repair rate $\mu(t)$ is given by:

$$\mu(t) = \frac{m(t)}{1 - M(t)} \tag{78}$$

Using similar arguments to the ones adopted for failure rate and reliability determination, repair time distribution function (maintainability) is given by:

$$M(t) = 1 - \exp\left[-\int_0^t \mu(\tau)\, d\,\tau\right] \tag{79}$$

and from Eqs. 79 and 78 the repair time probability density function is given by:

$$m(t) = \mu(t)\ \exp\left[-\int_0^t \mu(\tau)\, d\,\tau\right] \tag{80}$$

From the modeling point of view, repairable systems can be divided into the following two groups:

1. repairable systems for which failure is immediately detected (revealed failures);
2. repairable systems for which failure is only detected upon inspection (sometimes referred to as periodically inspected (tested) systems).

5.2 Repair of Revealed Failures

In systems with revealed failures the repair can be immediately initiated. In these situations two quantities are of primary interest, the number of failures over a given span of time and the system availability. Providing that the MTTR is much smaller than the MTTF, reasonable estimates for the number of failures can be obtained using the Poisson distribution and neglecting system downtime for repair. Availability may be calculated for constant repair rate and constant repair time.

Constant Repair Rates

In the case of a repairable component with a constant repair rate $\mu(t) = \mu$, the probability density function of the repair time will become exponential:

$$m(t) = \mu\ \exp\left(-\mu t\right) \tag{81}$$

and the mean time to repair:

$$\mathrm{MTTR} = \frac{1}{\mu} \tag{82}$$

Consider a system which may be operational or in failure. Availability $A(t)$ and unavailability $\tilde{A}(t)$ are complementary:

$$A(t) + \tilde{A}(t) = 1 \tag{83}$$

When the component has constant failure rate λ and repair rate μ, the availability variation between t and $t + \Delta t$ has a negative component related with the failure rate and a positive component resulting from the repair rate:

$$A\ (t + \Delta t) - A(t) = -\lambda\ \Delta t\, A(t) + \mu\ \Delta t\, \tilde{A}(t) \tag{84}$$

which leads to:

$$\frac{d}{dt} A(t) = -(\lambda + \mu)\, A(t) + \ \mu \tag{85}$$

Resolving the differential equation with an initial condition $A(0) = 1$, the instantaneous availability is given by:

$$A(t) = \frac{\mu}{\lambda + \mu} + \frac{\lambda}{\lambda + \mu} \exp[-(\lambda + \mu)t] \tag{86}$$

Note that availability is $A(t) = 1$ at $t = 0$ and decreases monotonically to an asymptotic value $1/(1 + \lambda/\mu)$, which depends only on the ratio of failure to repair rate.

The average or interval availability is obtained inserting Eq. 86 into 72:

$$\overline{A}(T) = \frac{\mu}{\lambda + \mu} + \frac{\lambda}{(\lambda + \mu)^2 T}\ [1 - \exp(-(\lambda + \mu)T)] \tag{87}$$

The asymptotic availability is then obtained when $T \rightarrow \infty$ in Eq. 87 (same value for instantaneous availability, i.e. $A(\infty) = A_\infty$) :

$$A_\infty = \frac{\mu}{\mu + \lambda} \approx 1 - \frac{\lambda}{\mu} \tag{88}$$

Since in most instants, repair rates are much larger than failure rates, a frequently used approximation comes from expansion and deleting the higher terms in λ/μ.

Equation 88 may be expressed in terms of mean time between failures and the mean time to repair:

$$A_\infty = \frac{\mathrm{MTTF}}{\mathrm{MTTF} + \mathrm{MTTR}} \tag{89}$$

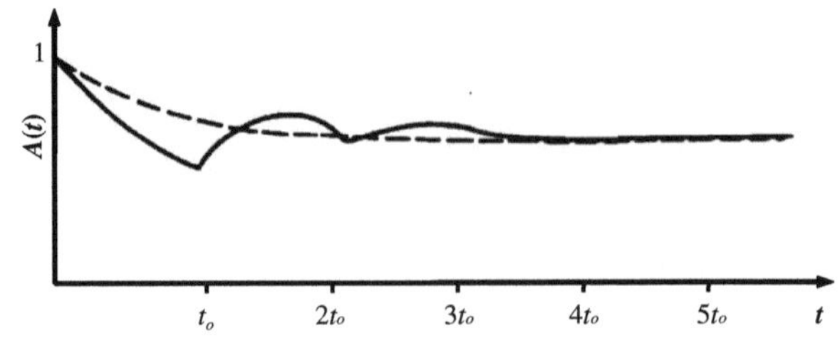

Fig. 9 Availability for constant repair time (*solid* line) and constant repair rate (*dashed* line)

Constant Repair Times

Considering now that all repairs have a constant duration t_o, the repair time density function is a Dirac delta function:

$$m(t) = \delta(t - t_o) \tag{90}$$

For time instants lower than t_o there are not any repairs and availability variation depends on the failure rate,

$$\frac{d}{dt}A(t) = -\lambda\, A(t) \quad \text{for } 0 \leq t \leq t_o \tag{91}$$

For times greater than t_o, repairs are also made; the number of repairs made during Δt is just equal to the number of failures during Δt at a time earlier $(t - t_o)$, thus the differential equation is

$$\frac{d}{dt}A(t) = -\lambda\, A(t) + \lambda\, A(t - t_o) \quad \text{for } t_o > t \tag{92}$$

During the first interval before t_o the solution of the differential equation is simply

$$A(t) = \exp(-\lambda t) \quad 0 \leq t \leq t_0 \tag{93}$$

For $t > t_o$, the solution in successive intervals, $Nt_o \leq t \leq (N+1)t_o$, depends on that of the preceding interval, therefore all intervals must be taken into account:

$$A(t) = A(Nt_o)\ \exp(-\lambda\,(t - Nt_o)) + \int_{N\,t_o}^{t} \lambda \exp(-\lambda(t-\tau))A(\tau - t_0)d\tau \tag{94}$$

or

$$A(t) = \sum_{i=0}^{N} \frac{\lambda(t - it_o)^i}{i!} \exp(-\lambda(t - it_o)) \text{ for } N\,t_o \leq t \leq (N+1)t_o \tag{95}$$

Figure 9 illustrates the availability for constant repair rate and constant repair time models. Note that the discrete repair time leads to breaks in the slop of the availability curve, whereas this is not the case with the constant failure rate model. However, both curves follow the same general trend downward and converge to the same asymptotic value. Therefore, the previous model is frequently adopted even for constant repair time situations.

5.3 *Repair: Unrevealed Failures*

For components in non-continuous operation, failures may occur but they are only detected when the component is put into operation. This kind of failures, occurring when the components are not in operation, can only be identified by *periodic testing*. However, the benefit of frequent testing in the detection of failures must be balanced with the loss of availability due to the downtime for testing and the possibility of excessive component wear from too frequent testing.

Idealized Periodic Tests

Starting by considering the effect of a simple periodic test on a system whose reliability can be characterized by a constant failure rate:

$$R(t) = \exp(-\lambda t) \tag{96}$$

The first thing that should be clear is that system testing has no positive effect on reliability. For unlike preventive maintenance the test will only catch failures after they occur. Testing may degrade equipment's reliability caused by additional wear from testing. Testing, however, has a very positive effect on system availability. To see this in the simplest case, suppose that a system test is performed at a time interval T_o. In addition the following assumptions are considered: (1) the time required to perform the test is negligible; (2) the time to perform repair is negligible; and (3) repairs are carried out perfectly and restore system to an as-good-as-new condition.

With the above conditions for a time period T_o, considering there is no repair, the availability is equal to the reliability:

$$A(t) = R(t) \quad \text{for } 0 \leq t < T_o \tag{97}$$

Since the system is repaired perfectly and restored to a as-good-as-new condition at $t = T_o$, the reliability will be $R(T_o) = 1$. Then since there is no repair between T_o and $2T_o$, the availability will again be equal to the reliability, but now the reliability is evaluated at $t - T_o$:

$$A(t) = R(t - T_o) \quad \text{for } T_o \leq t < 2T_o, \tag{98}$$

Leading to the general equation, represented in Fig. 10:

$$A(t) = R(t - iT_o) \quad \text{for } iT_o \leq t < (i + 1)T_o \tag{99}$$

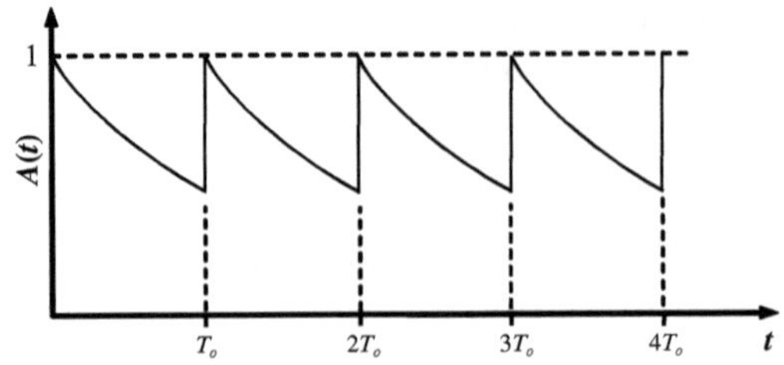

Fig. 10 Availability with idealized periodic testing for unrevealed failures

The interval and the asymptotic availability have the same value when the integral in Eq. 72 is taken over a multiple of T_o. Since the time interval is independent of the number of intervals over which the interval availability is calculated, so will the asymptotic availability A_∞ :

$$A_\infty = \lim_{n\to\infty} \frac{1}{nT_0} \int_0^{nT_0} A(t)\,dt = \frac{1}{T_0} \int_0^{T_0} A(t)\,dt \tag{100}$$

The effect of the testing interval is then obtained by combining Eqs. 100 and 96:

$$A_\infty = \frac{1}{\lambda T_0}(1 - \exp(-\lambda T_0)) \approx 1 - \frac{1}{2}\lambda T_0 \text{ when } \lambda T_0 \ll 1 \tag{101}$$

Real Periodic Tests

In the previous formulation derived for idealized periodic tests it has been assumed that the times required to perform the test and the repair are negligible. Based on these assumptions, it can be seen from Eq. 101 that it is possible to obtain availability near 1 by choosing small values to test interval T_o.

The hypothesis are now removed in order to represent the real situation in which test duration time T_t will lead to availability of zero during that time period. Maintaining the other hypothesis, from Eq. 101, the availability decreases to:

$$A_\infty \approx 1 - \frac{1}{2}\,\lambda T_0 - \frac{T_t}{T_0} \tag{102}$$

Considering now also the effect of the repair time characterized by a constant repair rate μ, the availability further decreases to:

$$A_\infty \approx 1 - \frac{1}{2}\,\lambda T_0 - \frac{T_t}{T_0} - \frac{\lambda}{\mu} \tag{103}$$

which shows that it is not sufficient to reduce the test interval T_o in order to affect availability, since it depends from T_t and μ as well, as shown in Fig. 11.

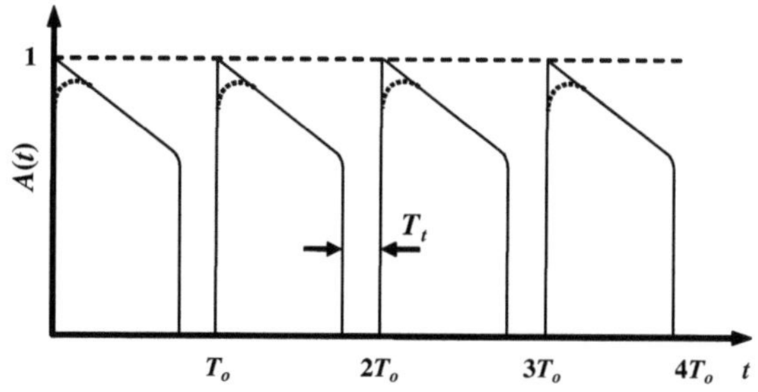

Fig. 11 Availability with realistic periodic testing for unrevealed failures

By differentiating Eq. 103 with respect to T_o and setting the result equal to zero, it is possible to determine the optimal test interval T_o^* (i.e. the one that maximizes the availability) by:

$$T_o^* = \sqrt{\frac{2T_t}{\lambda}} \tag{104}$$

The corresponding asymptotic availability is:

$$A_\infty^* = 1 - \sqrt{2\lambda T_t} - \frac{\lambda}{\mu} \tag{105}$$

5.4 System Availability

It is also important to examine the availability of a system in terms of the component availabilities. Not only are data more likely to be available at the component level, but that analysis can provide insight into the gains made through redundant configurations and through different testing and repair strategies.

For a series system with n independent components, system's availability is equal to the product of individual availabilities:

$$A(t) = \prod_i A_i(t) \tag{106}$$

where $A_i(t)$ is the availability for each component.

The parallel configuration requires all components to be unavailable is order for the system to be unavailable:

$$[1 - A(t)] = \prod_i [1 - A_i(t)] \tag{107}$$

Applying now the constant failure and repair rate models to each component, the asymptotic availability of a series system is

$$A_\infty = \prod_i \frac{\mu_i}{\mu_i + \lambda_i} \approx 1 - \sum_i \frac{\lambda_i}{\mu_i} \quad \text{if } \mu_i \gg \lambda_i \tag{108}$$

Considering a parallel system of n components, the asymptotic availability will be:

$$A_\infty = 1 - \left(\frac{\lambda_i}{\lambda_i + \mu_i}\right)^n \approx 1 - \left(\frac{\lambda_i}{\mu_i}\right)^n \quad \text{if } \mu_i \gg \lambda_i \tag{109}$$

6 Conclusions

This chapter has introduced the main concepts on reliability engineering building upon several textbooks selected as relevant in this subject area [1, 3, 4, 6]. However, this text and those textbooks deal mainly with concepts and methods and not their application to industrial problems, which some readers may be interested in.

For industrial applications a recent book [2] may be a good starting point to different industrial areas.

More advanced and new concepts on reliability engineering and applications can be found in scientific papers published in specialized journals, such as: Reliability Engineering & System Safety (RESS), IEEE Transactions on Reliability and Quality and Reliability Engineering International, among others.

References

1. Birolini A (2007) Reliability engineering theory and practice. Springer-Verlag, Berlin Heidelberg
2. Guedes Soares C (ed) (2010) Safety and reliability of industrial products, systems and structures. Taylor & Francis Group, London
3. Kumamoto H (2007) Satisfying safety goals by probabilistic risk assessment. Springer, Verlag London
4. Lewis EE (1996) Introduction to reliability engineering. Wiley, New York
5. Murthy P, Xie M, Jiang R (2004) Weibull models. Wiley Inc., New York
6. Rausand M, Hoyland A (2004) System reliability theory, models statistical methods and applications. Wiley, New York
7. Schneeweiss WG (1999) The fault tree method. LiLoLe-Verlag, Hagen

Fundamentals of Maintenance

Gilberto Francisco Martha de Souza and Fernando Jesus Guevara Carazas

Abstract The pieces of equipment installed in a thermal power plant must be subject to very stringent maintenance policies aiming at maintaining their operational availability. Those policies must be centered on controlling the aging degradation that affects mechanical equipment reliability. Furthermore the maintenance policy must account for economical and safety aspects associated with the failure of some critical equipment that constitutes the power plant. The use of Reliability-Centered Maintenance concepts is discussed as a suitable method to define equilibrium between the failure effects associated with the occurrence of critical equipment failure and the costs associated with maintenance policies applied to avoid the occurrence of those failures.

1 Introduction

In the broadest sense, reliability is associated with the successful operation of a device, with the absence of breakdowns or failures, for a given period of time, in a given set of operational conditions. The term reliability can be applied to almost any object, which is the reason that the terms system, equipment and component are used in the definition.

System is usually defined as a group of pieces of equipment assembled in a given functional configuration intending to perform a specific function. Reliability is defined positively, in terms of a system performing its intended function, and no

G. F. M. de Souza (✉) · F. J. G. Carazas
Department of Mechatronics and Mechanical Systems Engineering–Polytechnic School, University of São Paulo, Av. Prof. Mello Moraes 2231, São Paulo, 05508-900, Brazil
e-mail: gfmsouza@usp.br

G. F. M. de Souza (ed.), *Thermal Power Plant Performance Analysis*,
Springer Series in Reliability Engineering, DOI: 10.1007/978-1-4471-2309-5_6,

distinction is made between failures. Nevertheless, for system reliability analysis, there must be a great deal of concern not only with the probability of failure but also with the potential consequences of failures that present severe safety and economic loss or inconvenience.

During the system design stage the Failure Mode and Effects Analysis (FMEA) can be used to discover potential design weakness and enables the designer to rectify them at the design stage. Nevertheless, the components of a complex system deteriorate their performance during the operation and a maintenance policy should be specified in order to reduce the probability of failure of critical components during a given operational time.

The maintenance policy can be defined as a group of tasks that are executed in a system in order to assist the system in maintaining its functionality during operation.

One maintenance task can be defined as a set of activities which need to be performed, in specified manner, by the user in order to maintain the functionality of the system, Knezevic [6].

The number of activities and the type and quantity of resources required to maintain the system performance depend on the system design. The system inherent reliability will not be changed by any maintenance task.

The selection of a maintenance policy has great influence on the system operational cost and on its operational availability.

Until the middle 70s the maintenance focus was on trying to avoid failures using mainly scheduled maintenance programs. That maintenance philosophy can be applied to mechanical equipment where the failure mechanisms are usually well understood and the components design criteria provide information regarding the operational life. Based on that information the maintenance analyst developed the maintenance plan.

The applications of electronics in mechanical design aiming at improving control, automation and performance of machine increase the complexity of equipment. That complexity increased the difficult in understanding the possible failure modes presented by the components and how frequently would they occur.

The traditional scheduled maintenance policies were ineffective in controlling failure rates of those complex systems because for many items the probability of failure did not in fact increase with increasing operational age.

The aeronautical industry was the first sector that faced problems in defining maintenance program for airplanes with increasing complexity. The aeronautical industry changed the maintenance focus to keeping the system reliability and safety through the use of maintenance plans aiming at defining actions to control components reliability degradation and reducing maintenance costs in comparison with the traditional hard-time maintenance policy.

This development was the basis for the Reliability-Centered Maintenance (RCM) philosophy.

For RCM the driving element in maintenance decisions in not the way that a component failure but the consequences of that failure for the system operational condition and safety. The main goal of RCM is to develop an efficient scheduled-maintenance program subject to the constraints of satisfying safety and meeting operational performance goals.

This chapter presents the basic concepts of RCM philosophy aiming at its application in maintenance planning activities of thermal power plants.

2 Failure Modes of Complex Systems

Failures can be considered as inevitable in complex systems, although careful design, manufacturing and maintenance policy can control their occurrence and consequences. The design of complex equipment is always a trade-off between achieving the required performance and acceptable reliability. This problem enforces a compromise between the lightness and compactness required for high performance and the weight required for durability. Thus it is neither economically nor technologically feasible to produce complex equipment that can sustain trouble-free operation for an indefinite period of time. Furthermore, the use of electronics devices for monitoring and control of complex equipment aiming at increasing operational capability and safety also increased the variety of failures presented by equipment.

As for design and maintenance activities the term failure can be defined as the incapacity of a system, equipment or component to perform a function with a pre-defined performance.

Associated with the term failure two more definitions can be presented [10]:

1. Failure mode: is the effect by which a failure mode is observed;
2. Failure mechanism: is the chemical, physical or metallurgical process that leads to component failure.

The failure modes of a system depend on the functional arrangement of the components and on the nature of the components. Although those failure modes vary greatly due to the functional architecture, the initial failure events, the components failure modes, can be easily classified.

Electronic devices have basically three different failure modes: short circuit, open circuit and degraded performance. The failure mechanisms of electronic devices are not very well understood but usually involve electro mitigation, ion instability and others. The duration of the useful life of electronic components is very long and the failure rate is considered constant.

The components of mechanical and even electro-mechanical devices present failures usually associated with aging mechanisms or cumulative damage mechanisms, such as fatigue, wear and corrosion. The components usually fail when the damage exceeds the component's endurance level.

The failure rate curve for those components presents a long span of time with a monotonically increasing failure rate.

For both types of components it is possible to be identified an initial wear in period, as shown in Fig. 1.

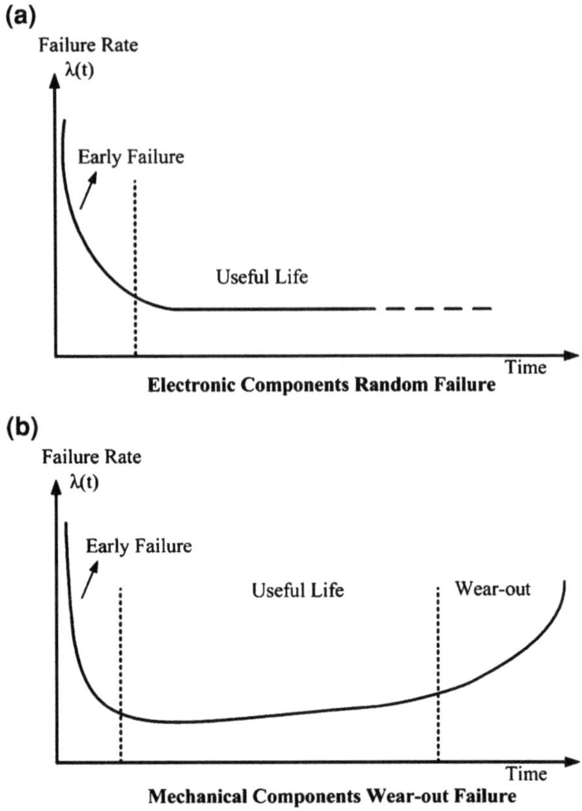

Fig. 1 Failure rate of a potential failure mode. **a** Electronic components random failure. **b** Mechanical components wear-out failure

Mechanical components may also present failures caused by overload. The failure modes associated with overloads can be presence of permanent distortion or fracture. Both failure modes affect the functionality, and consequently, the component reliability.

A summary of possible failure modes for mechanical components is presented in Tables 1 and 2.

3 Basic Maintenance Tasks

The maintenance tasks can be classified in three groups, which are: corrective, preventive and predictive (or condition-based) [14].

The corrective tasks are performed with the intention of restoring the functionality of the device after the loss of function or performance. The basic activities involved in corrective tasks are presented in Fig. 2. Depending on the nature of the device under analysis, the most time consuming activities are the failure diagnosis

Table 1 Summary of overload failure modes for mechanical components, adapted from Wang and Roush [16]

Failure mechanism	Description	Failure mode
Service limit state failure (Tensile-yield strength failure)	Occurs when the applied equivalent stress exceeds the yield strength of the material	Permanent deformation in the structural or mechanical component
Ultimate limit state failure (Ultimate tensile-strength failure)	Occurs when the applied equivalent stress exceeds the ultimate tensile strength	Causes rupture of the structural or mechanical component at a cross-sectional point
Compressive failure	Similar to the preceding tensile failures only under compressive loads	Permanent deformation or rupture
Failure due to shear loading	Occur when the shear stress exceeds the shear strength of the material when applying high torsion or shear loads	Yield (permanent deformation) or ultimate (cross section rupture)
Brittle fracture	Certain materials have little capability for plastic flow and are generally brittle, and thus are extremely susceptible to surface flaws and imperfections	Cracks propagate rapidly and completely through the component when the fracture stress is reached
Ductile fracture	High energy needed to continue the fracturing process	Relatively slow crack growth rates.
Instability failures (Buckling)	Instability failure occurs in structural components particularly those made from thin material and where the loading is generally in compression	A permanent deformation. Loss of mechanical stiffness
Bending failures	A combined failure where an outer surface is in tension and other surface is in compression	Permanent deformation or rupture of the structural or mechanical component

or disassembly and assembly. For electronic devices the failure detection can be very difficult demanding a lot of tests. For mechanical devices the failure detection is usually a very simple activity but depending on the failed component and on the device architecture the disassembly and assembly tasks can be very difficult.

A preventive maintenance task is performed in order to reduce the probability of failure of the device aiming at maximizing operational availability. The typical preventive maintenance tasks are presented in Fig. 3. The tasks are executed in a time based schedule aiming at avoiding the occurrence of components failures.

The most common preventive maintenance tasks are replacement and overhaul. Scheduled replacement is replacement of a device (or one of its parts) at or before some specified age limit. A scheduled replacement task is applicable only under the following circumstances: the item must be subject to a critical failure; test data must show that no failures are expected to occur below the specified life limit; the

Table 2 Summary of aging failure modes for mechanical components, adapted from Wang and Roush [16]

Failure mechanism	Description	Failure mode
Fatigue failure	Repeated loading and unloading of a component, with cycling load. The maximum stress can not cause tensile rupture.	Mechanical parts rupture if the design-life is exceeded.
Metallurgical failure	Caused by extreme oxidation or operation in corrosive environments.	Loss of stiffness. Possible rupture under normal loading condition.
Mechanical wear	Failures occur when surfaces moving in contact are damaged.	Higher friction and further damage.
Failure due to stress concentration	Failure occur due to an uneven stress, while stress concentrations take place at abrupt transitions from tick gauges to thin gauges, at abrupt changes in loading along a structure, at right-angle joints, or at various attachment conditions.	Failure at stress-concentration locations.
Failure due to flaws in materials	Failure due to improper inspection of materials, weld defects, fatigue cracks and other flaws.	Reduces the fatigue design-life. Mechanical parts rupture if the design life is exceeded.
Creep	Permanent elongation due to the formation of a network of micro-cracks due to temperature effects.	Permanent dimensional changes and deformation in metallic mechanical parts. Rupture if creep design-life is exceeded.

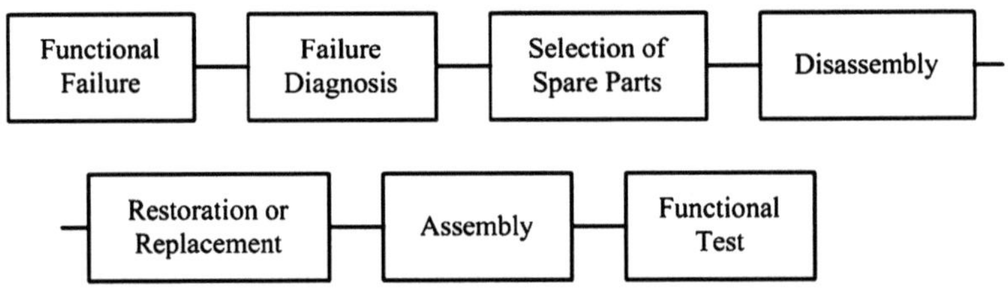

Fig. 2 Flowchart of activities of a corrective maintenance task, adapted from Knezevic [6]

item must be subject to a failure that has major economic (but not safety) consequences; there must be an identifiable age at which the item shows a rapid increase in the failure rate function and a large proportion of the units must survive to that age. A scheduled overhaul is a more complex scheduled replacement where a great number of devices or components of a system are replaced before some specified age limit.

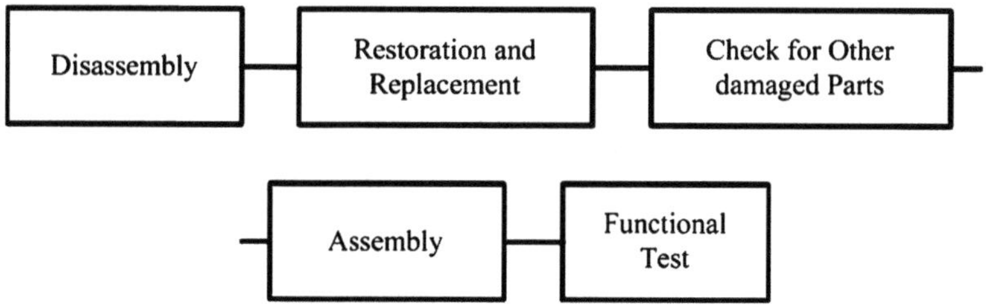

Fig. 3 Flowchart of activities of a preventive maintenance task, adapted from Knezevic [6]

Based on the discussion regarding components failure modes this type of maintenance is applied on mechanical devices that present wear out failures.

The need for provision of safety and reduction of maintenance costs has led to an increasing interest in development of alternative maintenance tasks.

In the last 20 years, with the great development of sensors and data acquisition and processing systems, the maintenance approach that seems to be the most attractive for minimizing the limitations of corrective or preventive practices is the predictive (or condition-based) maintenance task. This maintenance practice recognizes that the change in the condition and/or performance of a device is the most significant reason for carrying out maintenance and execution of preventive maintenance tasks should be based on actual condition of the device (or component). Thus, through monitoring some parameter(s) it should be possible to identify the most suitable instant of time at which preventive maintenance tasks should take place.

The condition maintenance task is based on condition monitoring activities which are performed in order to determine the physical state of a device/component of a system. Therefore, the aim of condition monitoring, whatever form it takes, is to monitor those parameters which provide information about the changes in condition and/or performance of a device/component that affect the system overall performance. The assessment of the current condition of a device/component is made by the use of techniques which can range from human sensing to sophisticated instrumentation, in order to determine the need for performing a preventive maintenance activity. The typical predictive maintenance tasks are presented in Fig. 4.

The condition assessment through condition monitoring of selected parameter(s) enables the maintenance planner to identify the most suitable instant of time at which preventive maintenance tasks should take place. So, the preventive maintenance tasks are not performed as long as the condition of the device/component is acceptable.

There are three criteria that must be met for an on-condition task to be applicable: it must be possible to detect reduced failure resistance for a specific failure mode; it must be possible to define a potential failure condition that can be detected by an explicit non-destructive inspection task; there must be a reasonable

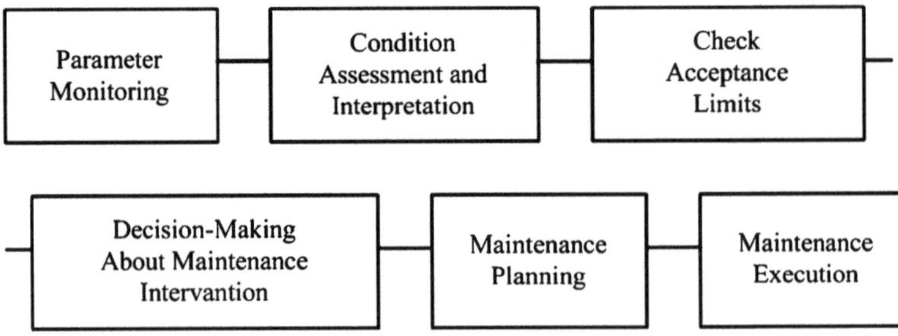

Fig. 4 Flowchart of activities of a predictive maintenance task, adapted from Knezevic [6]

Fig. 5 Development of a potential failure mode

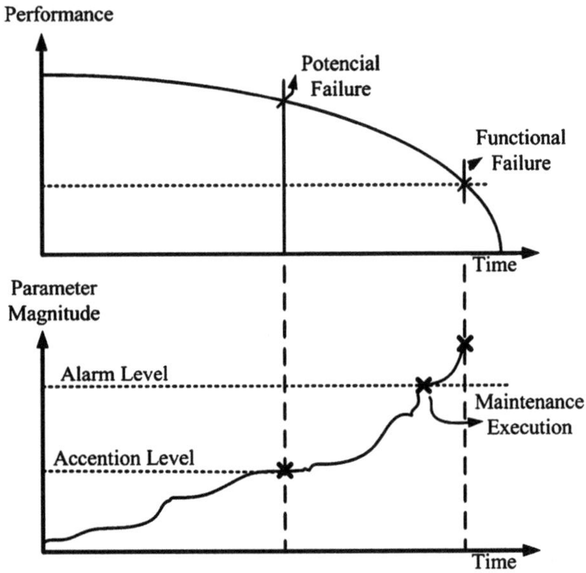

consistent age interval between the time of potential failure (P) and the time of functional failure (F), as illustrated in Fig. 5.

Condition monitoring devices are used to monitor, detect and diagnose the condition of the component/system monitored. Many techniques can be used for those activities aiming at providing information with respect to the actual condition of the component/system. Based on the analysis of the failure modes that can be presented by the component/system under analysis the most effective monitoring technique can be selected. The decision by which the condition monitoring techniques are chosen depends mainly on the type of component/system under analysis and, in the end, is driven by economic and/or safety decisions.

Effective condition-based maintenance tasks require a large number of measurements to be taken at specific time intervals that assure recognition of change in condition of the component/system in sufficient time for the execution of preventive action, demanding a great amount of data to be processed and analyzed aiming at diagnosing the failure development. The diagnostics for systems have now developed to a point where the information available is such that the analyst requires computer assistance to make the decision regarding the system operational state and maintenance intervention plan.

Some of the most frequent used condition monitoring techniques are [6]:

1. Vibration based on the fact that rotating machines such as pumps, compressors, turbines and electrical generators and motors produce vibration as they deteriorate. If those machines present the development of a failure mode their vibration levels and patterns change and vibration monitoring devices can be used to detect and to analyze these changes. Vibration monitoring equipment consists of three main items: transducers used for data acquisition, signal processing system and a algorithm by which the condition of the equipment being monitored is assessed. This algorithm is the basis for decision-making problem associated with the programming of any maintenance intervention before the occurrence of failure.
2. Lubrication oil analysis aiming at determining whether the fluid still meet the lubricating requirement. The results of the analysis is used to determine the lubricant life and to schedule oil changes intervals aiming at maintaining satisfactory equipment operation.
3. Wear particle analysis aiming at understanding the components/equipment wear processes and detecting abnormal system condition. Samples of lubricating oil are collected and analyzed in pre-defined time intervals and the change in the rate of debris collection indicates a change in the condition of the system.
4. Performance monitoring based on the use systems of sensors that monitor some performance parameters, sometimes referred as process parameters. Those indicative can be used to diagnose the system operational condition and a failure condition is recognized when certain limit values are exceeded. Those data are usually obtained with the use of thermometers, pressure gauges and flow meters.
5. Non-destructive techniques used to monitor the deterioration of some components, usually static equipment, such as pressure vessels, piping systems, and structural components. The basic advantage of the use of non-destructive technique to inspect static equipment is to evaluate the location and size of cracks close to hot spot stress areas. Using structural analysis methods it is possible to define the importance of the detected defects as for equipment structural integrity and to define the need for immediate maintenance action or for future maintenance. The non-destructive techniques include magnetic particle inspection, eddy current inspection, acoustic emission testing, radiographic inspection and ultrasonic inspection. Thermographic inspection is also considered as a non-destructive technique, but usually used to evaluate possible cumulative damage evolution in electrical equipment. Thermography uses

instrumentation designed to measure emissions of infrared energy as a mean of determining the operational condition of a piece of equipment. The most known use of thermography is in detecting failures in electrical circuits and in furnace walls.

4 Reliability-Centered Maintenance Concepts

According to Rausand [13], the maintenance plans usually used by industrial assets, including power plants, are often based on a combination of recommendations from manufacturers, legislation, company standards and maintenance models and data, the last one in a minor extent.

The main goal of the Reliability Centered Maintenance philosophy is to reduce the maintenance costs, centering the maintenance focus on the system functionality (on preserving the most important functions of the system as for company productivity and operational, environmental and personnel safety). To achieve this goal the RCM philosophy aims at avoiding or removing maintenance actions that are not necessary to keep the system functionality.

The main hypothesis that supports RCM application is that the component or equipment reliability is a function of the design and manufacturing process quality. The maintenance plan can only reduces the reliability decreasing during operational life but can not increase the design reliability.

The RCM analysis is executed through a sequence of steps, as shown in Fig. 6.

The main objective is to provide answers to the following questions:

1. What are the functions and associate performance standards of the components and equipment of a system in its present operating context?
2. In what way do the components or equipment fail to fulfill their functions (functional failures)?
3. What is the cause (physical mechanisms) of each failure mode that defines the functional failure?
4. What happens when each failure mode occurs?
5. In what way does each failure mode matter?
6. What can be done to prevent each failure mode?
7. What should be done if a preventive task cannot be proposed for a given failure mode?

The method first step involves the definition of the systems to be analyzed in order to improve maintenance planning.

In order to apply RCM philosophy the plant operator must identify the most important (or critical) components which failures affect the plant functions. Once the critical components are defined a maintenance policy can be proposed for them considering the RCM concepts.

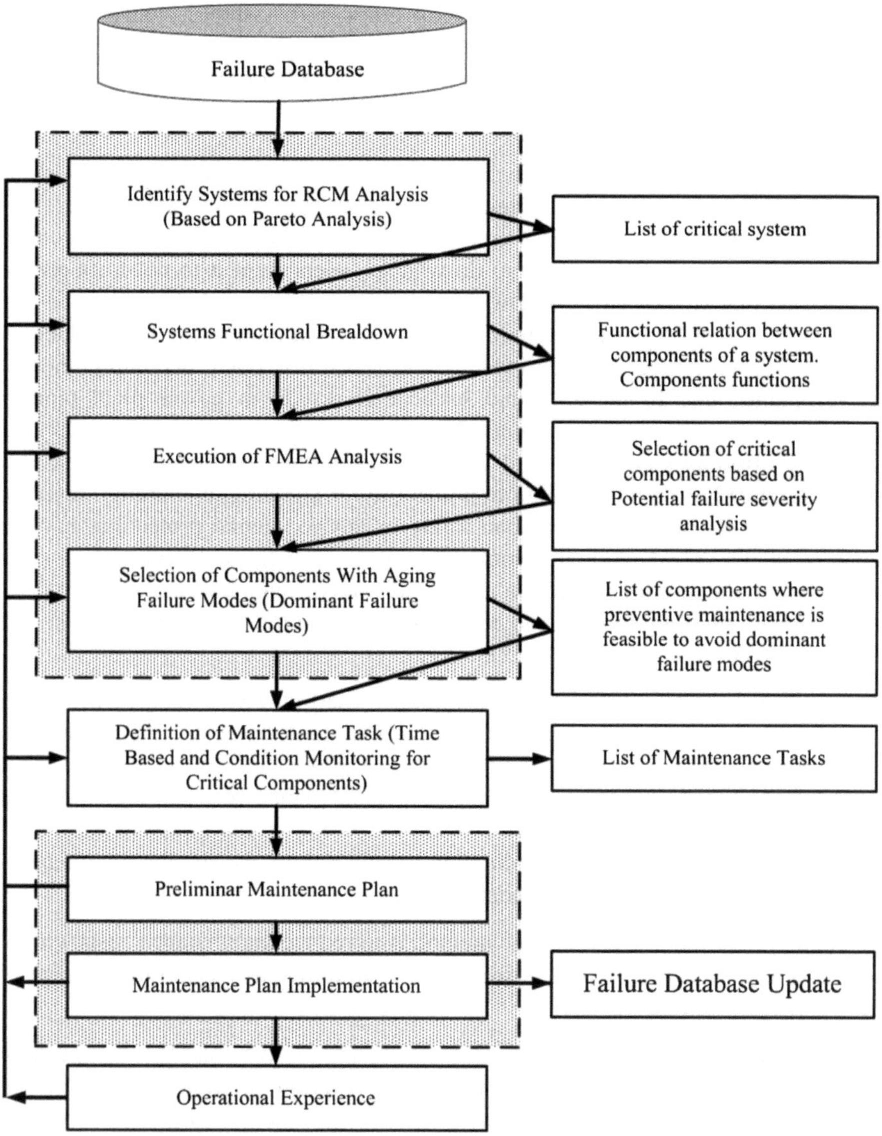

Fig. 6 Flowchart of RCM tasks

This maintenance policy philosophy is focused on the use of predictive or preventive maintenance tasks that aim at the reduction of unexpected failures during the component's normal operation, Moubray [11].

The RCM methodology was developed for the first time by United Airlines company for the American Department of Defense and was published in 1978,

Nowlan and Heap [12]. This, like several other initial studies, centered around challenging the traditional approach to scheduled maintenance programs which were based on the concept that every item on a piece of complex equipment has an age at which a complete overhaul is necessary to ensure safety and operating reliability.

Reliability Centered Maintenance is a continuous improvement framework for defining and optimizing the maintenance requirements for physical assets.

RCM is defined as "a process used to determine what must be done to ensure that any physical asset continues to do what its users want it to do in its present operating context…", SAE JA1011 standard for RCM, SAE [15].

The main idea of that maintenance policy selection philosophy is to reduce the amount of preventive interventions in any industrial plant, including thermal power plants, due to its costs and because it results in power plant planned unavailability, reducing the amount of produced energy.

The philosophy focuses on the definition of important equipment, equipment on which failures have a non-negligible probability of affecting availability, resulting in costly repairs or threats to human safety or environmental protection. For that equipment, the RCM method recommends, if technically and economically feasible, the application of predictive and/or preventive maintenance tasks.

To select such equipment and to define the required maintenance tasks, the RCM method recommends, Moubray [11]:

1. Analyze the causes and effects of failures, based on FMEA analysis, on plant availability, unavailability and corrective maintenance costs, plant safety and environmental degradation;
2. Analyze, based on operational data and maintenance records, the equipment reliability and maintainability;
3. Apply RCM decision-making diagram to select the most suitable maintenance practices for the equipment under analysis.

While the prime objective of the method is to reduce maintenance costs without affecting (or even improving) power plant availability and operational safety, it also enables, Moubray [11]:

1. Defining a well-justified predictive and/or preventive maintenance program which traceability will facilitate future evolution, aiming at continuously improving plant availability under a given set of operational conditions;
2. Making better use of operational feedback (including 'time to failure' and 'time to repair' database) which can be better structured and consequently can allow improvements in maintenance planning;
3. Adopting a functional approach to maintenance and linking it to the power plant operational equipment effectiveness.

The RCM method proposes a logical approach based on number of concepts described in the following sections.

Table 3 Failure mode and effects analysis form, adapted from Despujols et al. [3]

Component:				Subsystem		
Funtion:				System		
Functional failure	Potential failure mode	Mechanism of failure	Potential thermal power plant effect	Severity	Frequency of occurrence	Potential detection method

4.1 Functional Description

Initially the analyst must define the systems of the power plant which maintenance plan should be re-evaluated based on RCM concepts. Usually the application of that philosophy is recommended for the systems that execute the most important functions in a power plant. For a thermoelectric power plant those systems include the turbines (gas or steam) with their respective electric generators, steam generator, condenser, cooling towers, and water make-up.

The functions of the systems must be clearly identified. In order to perform a given function the system must have a group of components executing theirs functions according to a given pre-defined performance criteria.

In order to allow the functional analysis of the components of the system, the last one must be broken down into subsystems, sub subsystems, and so on, until the last element in the functional hierarchy, that is the component. The last component in the hierarchical functional structure is the one submitted to a maintenance task, including restoration or substitution in case of failure. The component can also be defined as an item that is able to perform at least one significant function as a stand-alone item, Rausand [13].

4.2 Failure Modes and Effects Analysis

Based on FMEA analysis, the analyst (or group of analysts) studies, for each piece of equipment (or system) installed in a thermal power plant, their components failure modes. For each component failure mode the severity, frequency or likelihood of occurrence and detection should be evaluated. For the detection analysis, the analyst must consider the possibility of monitoring failure evolution during plant operation. The authors suggest the use of Table 3 for FMEA analysis.

The analyst must determine the criticality of each failure mode. The criticality of a failure mode is the chance of loss or harmful effects that this mode generates per unit of time. Criticality is an indicator of the degree of concern one should have regarding a given failure mode. For maintenance planning, the higher the criticality the higher the level of effectiveness required of all actions performed to prevent the occurrence of the given failure mode.

Table 4 Relative severity criteria for hazardous events classification

Description		Set		
		Personal	Facilities	Environment
Insignificant	**I**	No significant harm to people, without removal of staff in the interior of the installation	No significant harm to installation	No significant harm to installation, contamination of environment in minimum concentration
Minor	**II**	Slight harm to people in installation, no significant harm to people outside installation	Minor damage or degradation of the installation, with repair at low cost	Contamination of environment below maximum concentration, though, concentration between minimum and medium
Major	**III**	Serious harm to people in installation and/or slight harm to people outside installation	Major damage or degradation of the installation, with possible repair	Contamination of environment below maximum concentration, though, concentration between medium and maximum
Catastrophic	**IV**	Single fatality or multiple severe harm to people inside and outside of installation	Damage or degradation without possible repair or repair take a long time to do	Contamination of environment above maximum concentration

The RCM criticality analysis is based on the evaluation of the component failure effect on the equipment or system under analysis (named severity) and on the frequency of occurrence of the failure. For the definition of the system degradation, the FMEA analysis usually uses a numerical code, usually ranking from 1 to 10. The higher the number the higher is the severity of the component that must be evaluated for each component failure mode. Carazas and Souza [1], when analyzing a gas turbine reliability, proposed as for severity analysis to classify that index in three main severity levels: marginal, critical and catastrophic. Each level is split into three other sub-levels to express some variety of failure effects. A severity scale between 1 and 9 is proposed. Values between 1 and 3 express minor effects on the turbine operation while values between 4 and 6 express significant effects on the turbine operation. Finally, failures that cause turbine unavailability or environmental degradation are classified by severity values between 7 and 9.

As for oil and gas industry, that could also be applied in the analysis of thermal power plants, Martins et al. [9] proposed the use of a severity matrix (Table 4) with four levels.

Operational feedback provides information on the frequency of failures of equipment components. The operational and maintenance records also provide information regarding:

Table 5 Frequency/probability for hazardous events classification

Description		Definition	
Extremely remote	**A**	More than 1 in 100.000 years	Extremely unlikely to occur during an activity, but should not exclude the existence of the hazard
Remote	**B**	More than 1 in 1,000 years	Highly unlikely to occur during an activity, but should not exclude the existence of the hazard
Less probable	**C**	More than 1 in 30 years	Possibly happened once during an activity
Probable	**D**	More than 1 in 10 years	Possibly happened more than once during an activity
Frequent	**E**	More than 1 in 1 year	Frequently happened during an activity

- The frequency of degradations observed during operation associated with in-service monitoring, inspections or tests.
- The frequency of degradations observed during preventive maintenance actions. These are degradations that have resulted in replacement of the components.

The frequency of failure can be also measured numerically, in a code varying from 1 to 10. The higher the level code the higher is the frequency of occurrence of failures of the component or equipment under analysis. Martins et al. [9] also presented a classification code as for frequency of failure analysis (Table 5).

The criticality can be defined based on a chart that classifies levels based on severity and frequency of failure of each component or equipment under analysis. In Fig. 7 the criticality matrix proposed by Martins et al. [9] as for system used in oil and gas industry is presented.

The criticality analysis of a component or equipment must consider all potential failure modes. A failure mode is a manner by which a failure is observed and is defined as a non-fulfillment of one of the functions of the component or equipment under analysis.

Additional information can be used to define the critical components. Other important information that can be added to the analysis is the maintenance costs of the pre-selected critical items. Those costs include the repair costs (that are associated with the equipment maintainability which influences the time required to disassembly and assembly the equipment during a maintenance action), lead time for spare parts and need for external personnel to execute maintenance actions.

Aiming at reducing power plant unavailability and controlling the plant overall maintenance costs, the RCM method recommends that the maintenance planning must focus primarily on the most critical equipment. For that set of equipment, maintenance tasks are selected with the application of maintenance decision-making diagram, as presented in Fig. 8, Nowlan and Heap [12].

In order to prepare the maintenance selection task the analyst must execute the reliability analysis of the critical components. The definition of the probability function that represents the reliability of a critical component allows the study of

		FREQUENT				
		A	B	C	D	E
SEVERETY	IV	MEDIUM	MEDIUM	HIGH	HIGH	HIGH
	III	MEDIUM	MEDIUM	MEDIUM	HIGH	HIGH
	II	LOW	LOW	MEDIUM	MEDIUM	MEDIUM
	I	LOW	LOW	LOW	LOW	MEDIUM

Fig. 7 Matrix for criticality selection

the dominant failure mode and its nature, defined by the failure rate. Based on the failure rate curve shape the analyst can evaluate the critical components which failure modes are based on aging effects and those which present random failure rate. This analysis supports the selection of maintenance tasks based on RCM philosophy.

The best source of information as for reliability analysis are the plant operational and maintenance database, where all information regarding components failure are stored, including operational time between failures and time to repair in case of failure, time to execute preventive tasks and other information associated to the cost of failure.

In case of lack of information about some component reliability, external files database where reliability information from systems with similar design and operating conditions are presented can be used to support preliminary reliability evaluation.

4.3 RCM Maintenance Task Selection Diagram

The RCM Decision Logic is used to determine what type of action would be appropriate to either eliminate or lessen the consequences of functional failures. Every function has one or more failure modes. Each of these failure modes must be processed through the Decision Logic to determinate if a preventive or predictive maintenance task can be developed to mitigate the consequences of their occurrence.

The Decision Logic requires that the following elements be considered for each critical component failure mode being analyzed:

- Consequences of failure (safety, environmental, operational, economical).
- Visibility of a functional failure to the operating crews.
- Age-reliability characteristics of each item.
- Economic trade-off decision based on a comparison of the cost of performing a preventive or predictive maintenance task to the cost of not performing the task.
- The Decision Logic consists of the four branches listed below:
- Evident Safety/Environmental consequences.

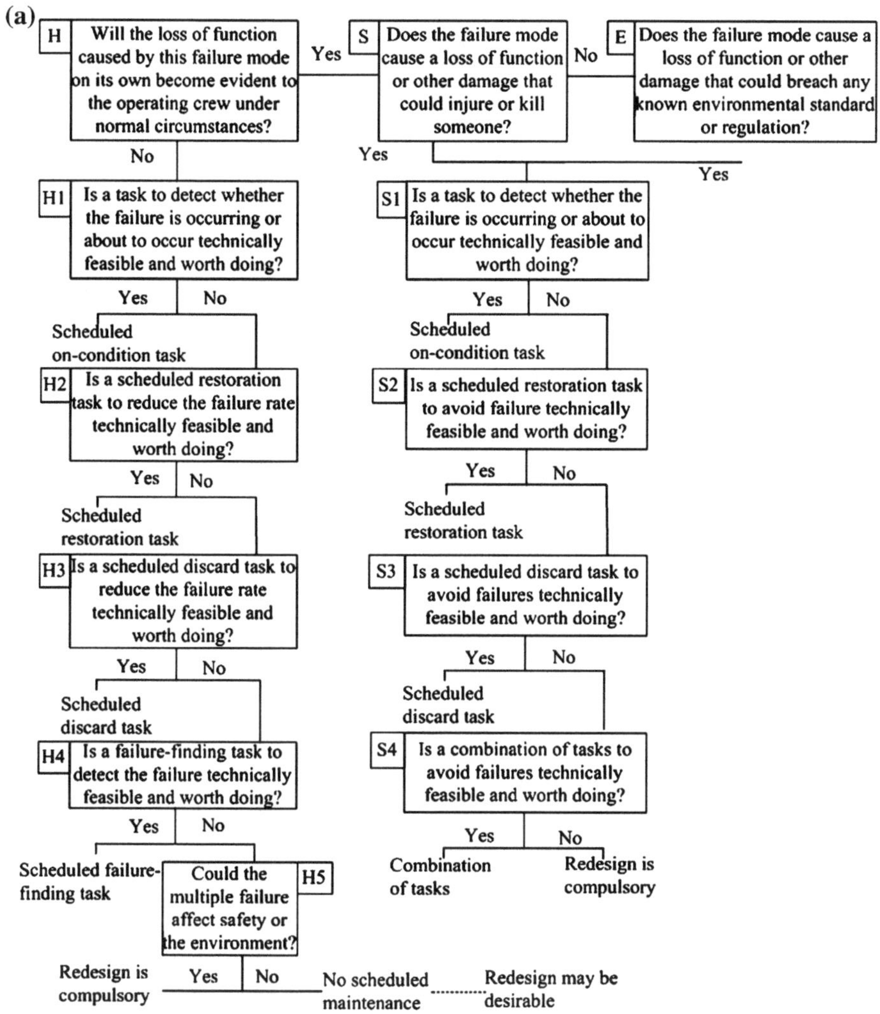

Fig. 8 RCM maintenance policy selection diagram. **a** Part 1, adapted from Nowlan and Heap [12]. **b** Part 2, adapted from Nowlan and Heap [12]

- Evident Economic/Operational Consequences.
- Hidden Economic/Operational/Safety/Environmental Consequences.

All four branches of the Decision Logic tree, shown in Fig. 8, propose the following two types of preventive maintenance tasks: on condition tasks (or predictive maintenance practice) and hard-time tasks. The Hidden Consequences contains proposal for failure finding tasks.

The first task found in the diagrams is use of predictive (or condition-based) maintenance including inspections and in-service monitoring. Those tasks aim at

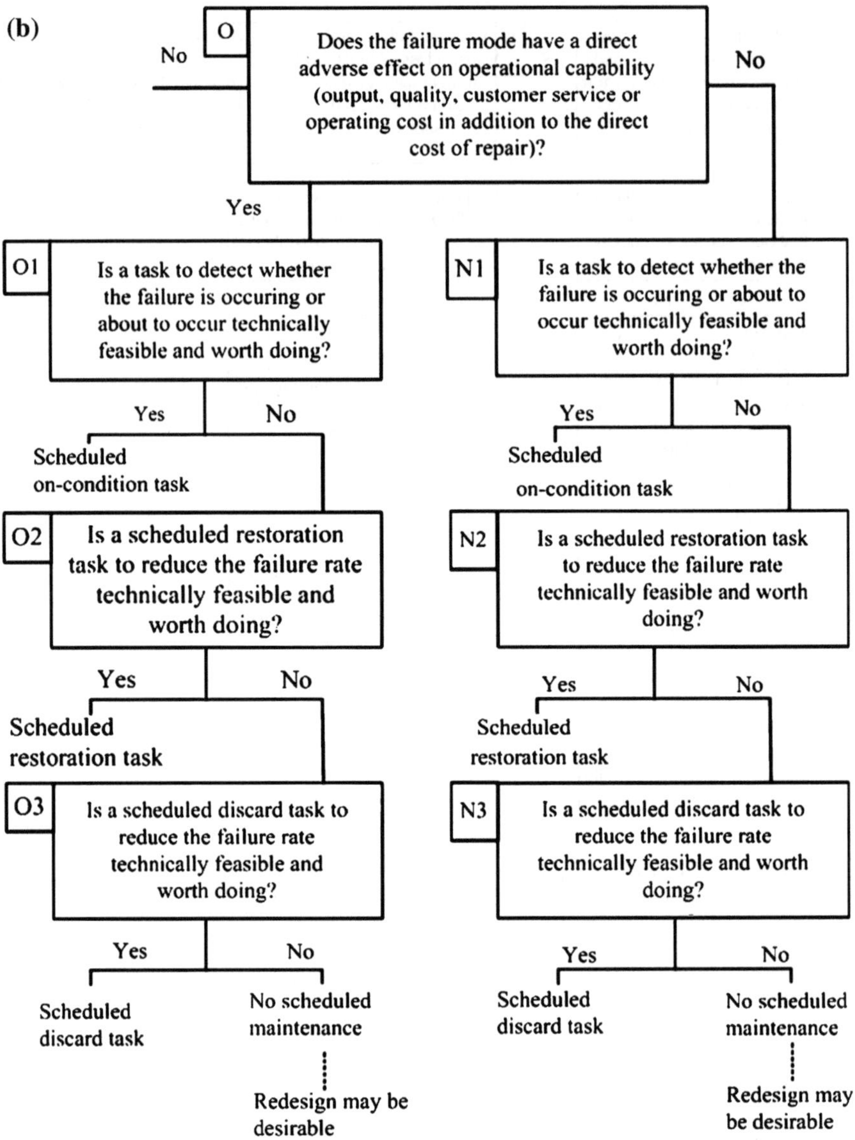

Fig. 8 (continued)

detecting degradations so as to perform condition-based repairs on the equipment. They are carried out while the equipment is in-service and therefore do not affect production. In case of detection of a potential failure that would lead to a functional failure the maintenance action can be planned in advance, including the necessary spare parts and external maintenance personnel. In case of maintenance action, the equipment must be disassembled and the damaged component substituted.

Scheduled replacements, representing preventive maintenance actions, are also recommended in the diagrams. The objective is to lower the failure rate of the component. That task usually demands shutdown and disassembly of the component, resulting in power plant planned shutdown according to a pre-defined schedule. These tasks are more efficient when the lifetime of the component is known.

A special case of scheduled replacements usually recommended for high cost power plant equipment, such as gas turbines or steam generators, is named scheduled overhaul. The overhaul represent a major repair in the equipment, aiming at revising and repairing (or substituting) all critical components that present an aging pattern failure rate. For example, heavy duty gas turbines are submitted to an overhaul before 48,000 operational hours, aiming at avoiding failures in the hot gas path.

The correct application of preventive maintenance tasks depends on the following hypothesis:

1. Reliability analysis must shown that the probability of failure during the specified operational time between two consecutive preventive intervention must be lower than a value defined by the operator;
2. The component must be subjected to a failure mode that has major economic consequences;
3. There must be an identifiable operational time at which the component shows a rapid increase in the failure rate.

For components that present functions that are required under some equipment specific operational conditions, a scheduled failure-finding task (named scheduled function test) of a hidden function can be applied. Those tasks are preventive only aiming at revealing failures of hidden functions. A scheduled function task is applied in a component that can present a functional failure that is not evident to the operator when the equipment is performing its normal duties.

For components which failure affect the operational and economical plant performance but do not affect safety and environment, the corrective maintenance task can be considered as an option if the failure mode presents a constant failure rate, being classified as a random failure.

For failure modes which occurrence can affect the operational safety or the environment and present constant failure rate, the use of preventive or predictive maintenance tasks will not reduce the probability of failure occurrence at an acceptable level. The components presenting that type of failure modes must be redesigned or modified aiming at increasing their reliability. Another option is the use of redundant component (usually in the stand-by mode), but that choice is dependent on a cost-benefit assessment.

One of the most important tasks after the definition of each critical component maintenance policy based on RCM decision diagram is the selection of the preventive maintenance intervals. According to [Rausand [13]] the determination of an optimum interval has to be based on information about the failure rate function and the likely consequences the PM task is supposed to prevent.

Lewis [7] and Duffua et al. [4] present a model to define the most suitable age replacement interval of components subject to preventive maintenance. The model is based on the minimization of maintenance cost of the component considering known the following data:

- R(t) reliability distribution for the component under analysis
- C_p cost of planned replacement, including labor costs and cost of the new item
- C_d cost of failure in service, including cost of new item, labor costs, loss of production costs, costs of any temporary provisions that have to be made as a result of the failure.

Supposing that the component will be replaced at time intervals of T units, even if a unexpected failure has occurred during operational interval T, the cost of maintenance for the component over the period T is the sum of:

- A fixed part, C_p, incurred at the end of the period
- A variable part nC_d, where n is the number of failures occurring during the period T.

The second part is a random variable whose average value is dependent on the probability of occurrence of n failures in the time interval T. Considering that T is sufficiently short, the average preventive maintenance cost of a component is presented in Eq. 1 [8]:

$$E(C) = C_p + [1 - R(T)].C_d \quad (1)$$

The selected preventive maintenance interval T is the one that minimizes the component average maintenance cost.

Once the preventive replacement time interval for each critical component has been calculated, the various maintenance tasks must be conciliated in packages that are carried out at the same time (for a given equipment or for a group of pieces of equipment) or in a pre-defined sequence. The maintenance intervals can not be optimized for each single component but it has to be treated as a group of tasks that must be analyzed as a whole.

Most of the users of RCM instead of searching for the optimum interval as for preventive maintenance try to reconcile the preventive and predictive recommendations with the existing maintenance program, which is usually based on manufacturers' recommendations and on past experience.

The decision diagram of RCM overrides two criteria for selecting maintenance. Each task selected for each critical component must be analyzed according to two requirements: applicability and cost-effectiveness.

The applicability is related to the possibility of elimination, or at least reduce the probability of occurrence to a given target level, of a specific failure mode.

The cost-effectiveness is a measure of how well the selected maintenance task accomplishes the purpose of applicability in relation to the cost of failure.

The cost-effectiveness analysis is based on the application of decision-making theory methods aiming at balancing the cost of performing the maintenance with the cost of not performing, leaving the component to run-to-failure. The run-to-failure maintenance policy is not feasible for the cases of failure modes that affect plant or environment safety.

As for cost analysis regarding the application of preventive or predictive maintenance tasks, the analyst must include the following items:

- The cost related to possible maintenance induced failures caused by errors in disassembly and assembly of equipment
- Production unavailability during maintenance
- Unavailability of protective functions during maintenance of their components

The cost of failure may include:

- The cost of failure consequences (loss of production, possible violations of regulations, damage to other equipment)
- Cost of emergency repairs (including spare parts acquisition and extra personnel necessary for repairs)

Taking in view that the maintenance task selection is based on risk, the decision tree application is recommended to support that selection.

4.4 Streamlined Reliability-Centered Maintenance

Although the RCM philosophy is considered as an important improvement in maintenance planning, once it focuses in critical components whose failures have important consequence on the system performance, regarding economic or safety aspects its application is also considered a time consuming task.

The selection of critical components based on the concern with maintaining system functionality through the application of FMEA for all components of equipment is considered the most difficult and time consuming analysis during RCM analysis.

Taking in view the possibility of lowering the economical efforts associated with the application of RCM concepts without affecting the technical aspects, EPRI [5] developed a process named Streamlined Reliability-Centered Maintenance (SRCM), considering their experience in applying RCM concepts in nuclear power plant systems.

The SRCM was developed as a project aiming at investigating possible methods of lowering costs to perform a RCM analysis while maintaining the technical integrity of the process and results.

The steps used in SRCM analysis are similar to those presented for the traditional RCM application. Nevertheless, some of the steps must be developed in a different way.

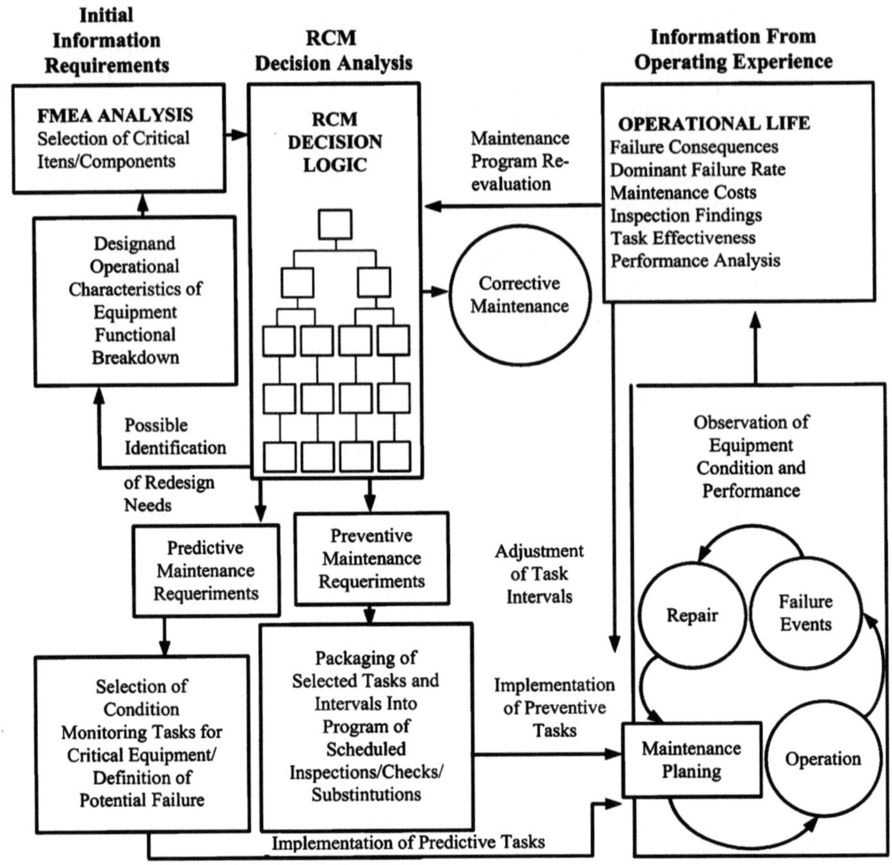

Fig. 9 RCM update process during piece of equipment operational life cycle, adapted from DoD [2]

The functional analysis of the plant pieces of equipment must be executed in a simplified way, dividing their functions in two categories: important functions and non-important functions. Important functions are the ones that directly affect plant safety, including environmental limits and power production.

Based on critical function selection, for each piece of equipment, the FMEA analysis is executed only for the components that support the critical functions while in the traditional RCM format each component must be evaluated in order to define their failure modes consequence on the plant operational condition.

So, in the SRCM process, the analyst identifies every component that supports the functional failure and lists only the most significant failure modes for each component along with the most dominant plant effects for the failure modes. The analyst determines the component criticality based on various failure mode/plant effect combinations and the cumulative significance of the components failure on a specific function based on a criterion defined by the organization.

If a component is determined to be critical, the standard RCM maintenance decision tree analysis can be used. To support that application, the analyst must identify appropriate causes for the failure modes aiming at identifying applicable and effective maintenance tasks to reduce the frequency of occurrence of those failure modes.

For non-critical equipment the analyst must select maintenance policies based mainly on economic aspects although emphasizing the application of preventive maintenance concepts.

5 Conclusion

Although the application of RCM philosophy to improve the maintenance plan of power plant equipment seems economically expensive, the advantages associated with that application as for power plant performance seems also very attractive.

RCM philosophy is an approach to maintenance that combines corrective, preventive and predictive practices to maximize the life of a piece of equipment aiming at performing a given function according to their inherent reliability capabilities.

The driving element in RCM decisions is not the failure of a given component but the consequences of that failure for the equipment or system as a whole. The principles of reliability-centered maintenance stem from a rigorous examination of certain questions that are of taken for granted?

1. How does a failure occur?
2. What are its consequences?
3. What good can predictive/preventive maintenance do?

Based on the answers for the previous questions it is possible to develop an efficient condition monitoring program or scheduled-maintenance program subjected to the constraints of satisfying requirements and meeting operational-performance goals.

Nevertheless, due to the initial investment required to obtain the technological tolls, training and plant equipment condition baselines, the implementation of a RCM program will typically result in a short-tem increase in maintenance costs. The increase is rapidly compensated by the decrease of corrective maintenance costs due to reduction of unexpected failures. The long-term maintenance costs will also be reduced by the replacement of hard time preventive tasks by predictive maintenance application.

Another advantage of RCM is that decision about equipment replacement is based on equipment condition, defined by condition monitoring database analysis, and not on calendar. This condition-based approach to maintenance extends the life of the facility and its equipment.

The application of RCM decision diagram requires the use of a decision-making technique to evaluate the feasibility of use of a given preventive or predictive maintenance task taking in view the economic impacts on the plant life cycle cost.

Once the RCM-based maintenance plan is implemented, a continuous monitoring of the maintenance actions and registration of failure occurrence must be executed. For each significant failure that occurs in the system, the failure characteristics should be compared with the FMEA analysis and, if necessary, the RCM-based maintenance plan must be revised.

Figure 9 summarizes the RCM analysis process throughout a piece of equipment operational life cycle.

References

1. Carazas FJG, Souza GFM (2009) Availability analysis of gas turbines used in power plants. Int J Thermodyn 12:28–37
2. Department of Defense (DoD) (1986) MIL-STD-2173 Reliability-centered maintenance for naval aircraft weapons systems and support equipment. Department of Defense, Washington
3. Despujols A, Delbos JP, Meilland JP, Zuliani G Reliability centered maintenance applied to gas turbines. Eletricite de France [in press]
4. Duffuaa SO, Raouf A, Campbell JD (1999) Planning and control of maintenance systems: modeling and analysis. Wiley, New York
5. Electric Power Research Institute (EPRI) (1998) TR-109795 Streamlined reliability-centered maintenance (srcm) program for fossil-fired power plants. EPRI, Palo Alto
6. Knezevic J (1997) Systems maintainability: analysis engineering and management. Chapman and Hall, London
7. Lewis E (1987) Introduction to reliability engineering. Wiley, New York
8. Lyonnet P (1991) Maintenance Planning: methods and mathematics. Chapman and Hall, London
9. Martins MR, Souza GFM, Rodriguez CEP (2009) Risk-based analysis of offloading operations with fpso production units. Proceeding of the 20th international congress of mechanical engineering (COBEM 2009), Gramado, RS, Brazil, p 10
10. Modarres M (1993) What every engineer should known about reliability and risk analysis. Marcel Dekker Inc., New York
11. Moubray J (1997) Reliability-centered maintenance. Industrial Press Inc., New York
12. Nowlan FS, Heap HF (1978) Reliability centered maintenance. Office of Assistant Secretary of Defense, Washington
13. Rausand M (1998) Reliability centered maintenance. Reliab Eng Syst Saf 60:121–132
14. Smith A, Hinchcliffe G (2004) Reliability-centered maintenance: a gateway to world class maintenance. Elsevier Butterworth-Heinemann, New York
15. Society of Automotive Engineers (SAE) (1999) SAE Standard JA1011 Evaluation criteria for reliability centered maintenance (RCM) process. SAE international, Warrendale
16. Wang JX, Roush ML (2000) What every engineer should know about risk engineering and management. Marcel Dekker Inc., New York

Fundamentals of Risk Analysis

Bilal M. Ayyub

Abstract This chapter presents the basic concepts associated with risk analysis of complex systems. The chapter provides definitions and terminology. It discussed quantification of risk in a systems framework. Various methods for risk assessment are discussed, and the risk management and control philosophies are presented. Risk-informed decision making is introduced on the basis of benefit-to-cost analysis. Methods for defining risk acceptance thresholds are provided for various system types from several industries. Also, risk communication methods are briefly described. The concepts presented are suitable for power plant performance improvement and safety. They can be adapted for power plant related applications. Sources are provides at the end of chapter for additional reading and details on risk methods.

1 Introduction

Risk is associated with all projects and business ventures taken by individuals and organizations regardless of their sizes, their natures, and their time and place of execution and utilization. The chapter defines and discusses risk and its dimensions, risk analysis, risk management and control, and risk communication, and is based on Ayyub [3] and other chapters prepared by the authors for other publications.

B. M. Ayyub (✉)
Center for Technology and Systems Management,
Department of Civil and Environmental Engineering,
University of Maryland, College Park, MD 20742, USA
e-mail: ba@umd.edu

G. F. M. de Souza (ed.), *Thermal Power Plant Performance Analysis*,
Springer Series in Reliability Engineering, DOI: 10.1007/978-1-4471-2309-5_7,

Definitions that are needed for presenting risk-based technology methods and analytical tools are presented in this section.

1.1 Hazards

A hazard is an act or phenomenon posing potential harm to some person(s) or thing(s), i.e., a source of harm, and its potential consequences. For example, uncontrolled fire is a hazard, water can be a hazard, and strong wind is a hazard. In order for the hazard to cause harm, it needs to interact with person(s) or thing(s) in a harmful manner. The magnitude of the hazard is the amount of harm that might result, including the seriousness and the exposure levels of people and the environment. Hazards need to be identified and considered in projects' lifecycle analyses since they could pose threats and could lead to project failures. The interaction between a person (or a system) and a hazard can be voluntary or involuntary.

1.2 Reliability and Performance

Reliability can be defined for a system or a component as its ability to fulfill its design functions under designated operating or environmental conditions for a specified time period. This ability is commonly measured using probabilities. Reliability is, therefore, the occurrence probability of the complementary event to failure as provided in the following expression:

$$\text{Reliability} = 1 - \text{ Failure Probability} \tag{1}$$

The performance of a system or component can be defined as its ability to meet functional requirements. The performance of an item can be described by various elements including such items as speed, power, reliability, capability, efficiency, and maintainability.

1.3 Event Consequences

For an event of failure, consequences can be defined as the degree of damage or loss from some failure. Each failure of a system has some consequence(s). A failure could cause economic damage, environmental damage, injury or loss of human life, or other possible events. Consequences need to be quantified in terms of failure consequence severities using relative or absolute measures for various consequence types to facilitate risk analysis.

1.4 Risk

The concept of risk can be linked to uncertainties associated with events. Within the context of projects, risk is commonly associated with an uncertain event or condition that, if it occurs, has a positive or a negative effect on project's objectives.

Formally, risk can be defined as the potential of losses and rewards resulting from an exposure to a hazard or as a result of a risk event. Risk should be based on identified risk events or event scenarios. Risk can be viewed to be a multi-dimensional quantity that includes event-occurrence probability, event-occurrence consequences, consequence significance, and the population at risk; however, it is commonly measured as a pair of the probability of occurrence of an event, and the outcomes or consequences associated with the event's occurrence. This pairing can be represented by the following equation:

$$Risk \equiv [(p_1, c_1), (p_2, c_2), \ldots, (p_i, c_i), \ldots, (p_n, c_n)] \tag{2}$$

where p_i is the occurrence probability of an outcome or event i out of n possible events, and c_i is the occurrence consequences or outcomes of the event. A generalized definition of risk can be expressed as

$$Risk \equiv [(l_1, o_1, u_1, cs_1, po_1), (l_2, o_2, u_2, cs_2, po_2), \ldots, (l_n, o_n, u_n, cs_n, po_n)] \tag{3}$$

where l is likelihood, o is outcome, u is utility (or significance), cs is causal scenario, po is population affected by the outcome, and n is the number of outcomes. The definition according to Eq. 3 covers all attributes measured in risk assessment that are described in this chapter, and offers a complete description of risk, from the causing event to the affected population and consequences. The population-size effect should be considered in risk studies since society responds differently for risks associated with a large population in comparison to a small population. For example, a fatality rate of 1 in 100,000 per event for an affected population of 10 results in an expected fatality of 10^{-4} per event whereas the same fatality rate per event for an affected population of 10,000,000 results in an expected fatality of 100 per event. Although, the impact of the two scenarios might be the same on the society (same risk value), the total number of fatalities per event/accident is a factor in risk acceptance. Plane travel may be "safer" than for example recreational boating, but 200–300 injuries per accident are less acceptable to society. Therefore, the size of the population at risk and the number of fatalities per event should be considered as factors in setting acceptable risk.

Risk is commonly evaluated as the product of likelihood of occurrence and the impact severity of occurrence of the event:

$$Risk\left(\frac{Consequence}{Time}\right) = Likelihood\left(\frac{Event}{Time}\right) \times \text{Impact}\left(\frac{Consequence}{Event}\right) \tag{4}$$

Fig. 1 Example risk profile

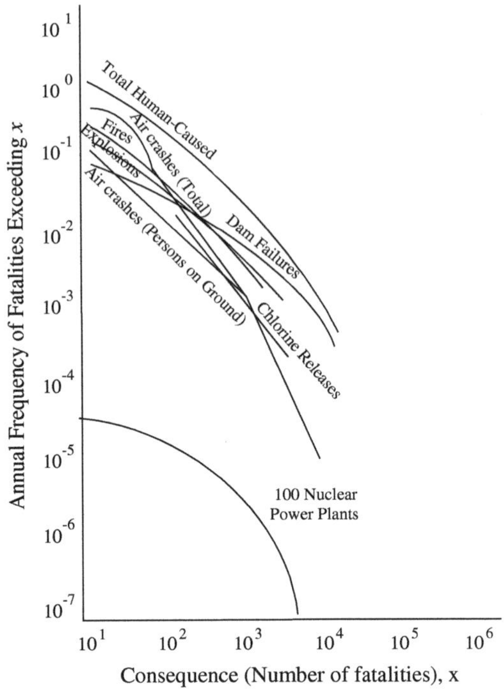

In Eq. 4, the likelihood can be measured in terms of a probability. Equation 4 presents risk as an expected value of loss or an average loss. A plot of occurrence probabilities and consequences is called a risk profile or a Farmer curve. An example Farmer curve is given in Fig. 1 based on a nuclear case study, provided herein for illustration purposes. It should be noted that the abscissa provides the number of fatalities, and the ordinate provides the annual frequency of exceedence for the corresponding number of fatalities. These curves are sometimes constructed using probabilities instead of frequencies. The curves represent or median average values. Sometimes, bands or ranges are provided to represent uncertainty in these curves. They represent confidence intervals for the average curve or for the risk curve. Figure 2 shows examples curves with uncertainty bands. This uncertainty is sometimes called meta-uncertainty. A complete treatment of uncertainty analysis is provided by Ayyub and Klir [4].

The occurrence probability (p) of an outcome (o) can be decomposed into an occurrence probability of an event or threat (t), and the outcome-occurrence probability given the occurrence of the event ($o|t$). The occurrence probability of an outcome can be expressed as follows using conditional probability concepts:

$$p(o) = p(t)p(o|t) \tag{5}$$

In this context, threat is defined as a hazard or the capability and intention of an adversary to undertake actions that are detrimental to a system or an organization's

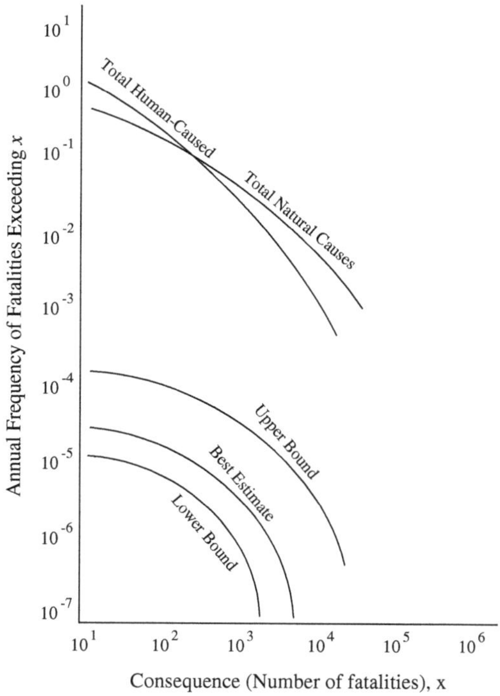

Fig. 2 Uncertain risk profile

interest. In this case, threat is a function of only the adversary or competitor, and usually cannot be controlled by the owner or user of the system. However, the adversary's intention to exploit his capability may be encouraged by vulnerability of the system or discouraged by an owner's countermeasures. The probability $[p(o|t)]$ can be interpreted as the vulnerability of the system in case of this threat occurrence. Vulnerability is a result of any weakness in the system or countermeasure that can be exploited by an adversary or competitor to cause damage to the system.

1.5 Risk-Based Technology

Risk-based technologies (RBT) are methods or tools and processes used to assess and manage the risks of a component or system. RBT methods can be classified into risk management that includes risk assessment/risk analysis and risk control using failure prevention and consequence mitigation, and risk communication as shown in Fig. 3.

Risk assessment consists of hazard identification, event-probability assessment, and consequence assessment. Risk control requires the definition of acceptable risk and comparative evaluation of options and/or alternatives through monitoring and

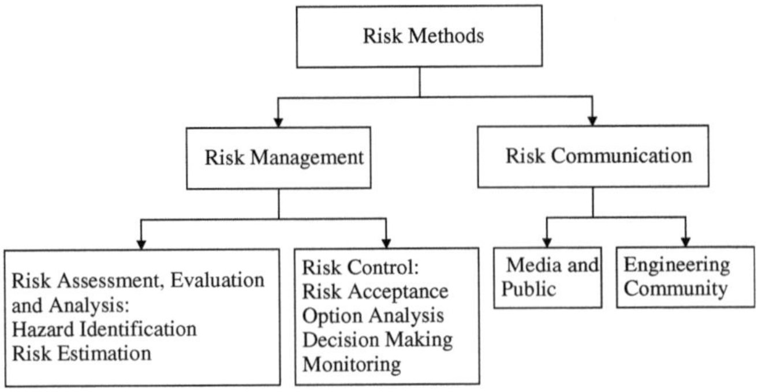

Fig. 3 Risk-based technology methods

decision analysis. Risk control also includes failure prevention and consequence mitigation. Risk communication involves perceptions of risk, which depends on the audience targeted, hence, classified into risk communication to the media, the public, and to the engineering community.

1.6 Safety

Safety can be defined as the judgment of risk acceptability for the system. Safety is a relative term since the decision of risk acceptance may vary depending on the individual making the judgment. Different people are willing to accept different risks as demonstrated by different factors such as location, method or system type, occupation, and lifestyle. The selection of these different activities demonstrates an individual's safety preference despite a wide range of risk values. Table 1 identifies varying annual risks for different activities based on typical exposure times for these activities. Also Fig. 4 from the Imperial Chemical Industries, Limited shows the variation of risk exposure during a typical day that starts by waking up in the morning from sleep and getting ready to go to work, then commuting and working during morning hours, followed by a lunch break, then additional work hours followed by commuting back to having dinner, and round trip on a motorcycles to a local pub. The ordinate in this figure is the fatal accident frequency rate (FAFR) with a FAFR of 1.0 corresponding to one fatality in 11,415 years, or 87.6 fatalities per one million years. The figure is based on an average number of deaths in 10^8 h of exposure to a particular activity.

Risk perceptions of safety may not reflect the actual level of risk in some activity. Table 2 shows the differences in risk perception by three groups of the league of women voters, college students, and experts of 29 risk items. Only the top items are listed in the table. Risk associated with nuclear power was ranked as the highest type by women voters and college students, whereas it was placed as the 20th by experts. Experts place motor vehicles as the first risk. Public perception

Table 1 Relative risk of different activities

Risk of death	Occupation	Lifestyle	Accidents/ recreation	Environmental risk
1 in 100	Stunt-person			
1 in 1,000	Racecar driver	Smoking (one pack/day)	Skydiving Rock climbing Snowmobile	
1 in 10,000	Fire fighter Miner Farmer Police officer	Heavy drinking	Canoeing Automobile All home accidents Frequent air travel	
1 in 100,000	Truck driver Engineer Banker Insurance agent	Using contra ceptive pills Light drinking	Skiing Home fire	Substance in drinking water Living downstream of a dam
1 in 1,000,000		Diagnostic X-rays Smallpox vaccination (per occasion)	Fishing Poisoning Occasional air travel (one flight per year)	Natural background radiation Living at the boundary of a nuclear power
1 in 10,000,000		Eating charcoal-broiled steak (once a week)		Hurricane Tornado Lightning Animal bite or insect sting

of risk and safety varies by age, gender, education, attitudes, and culture among other factors. Individuals sometimes do not recognize uncertainties associated with risk event or activity causing an unwarranted confidence in an individual's perception of risk or safety. Rare causes of death are often overestimated and common causes of death are often underestimated. Perceived risk is often biased by the familiarity of the hazard. The significance or the impact of safety perceptions stems from that decisions are often made on subjective judgments. If the judgments hold misconceptions about reality, this bias affects the decision. For example, the choice of a transportation mode—train, automobile, motorcycle, bus, bicycle, etc.—results in a decision based on many criteria including such items as cost, speed, convenience, and safety. The weight and evaluation of the decision criteria in selecting a mode of transportation rely on the individual's perception of safety that may deviate sometimes significantly from the actual values of risks. Understanding these differences in risk and safety perceptions is vital to performing risk management decisions and risk communications as provided in subsequent sections on risk management and control.

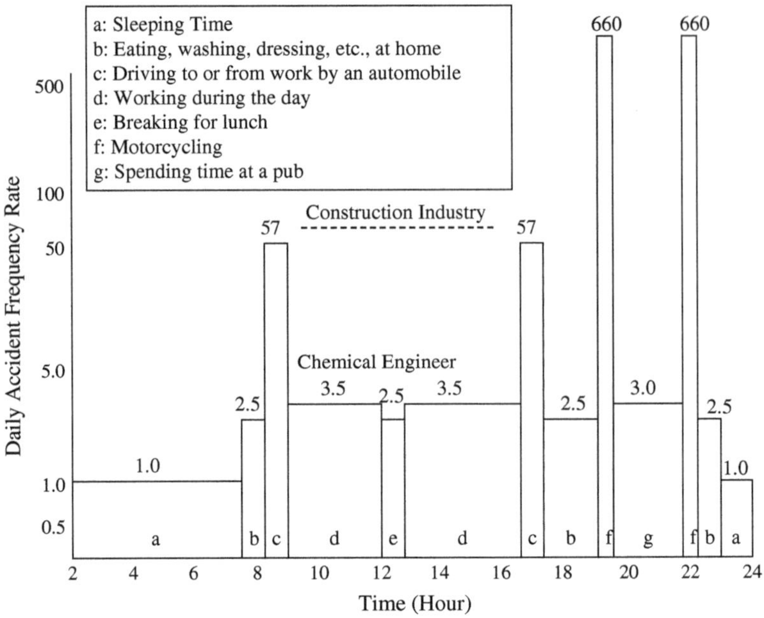

Fig. 4 Daily death risk exposure for a working healthy adult

2 Risk Assessment

2.1 Risk Assessment Methodologies

Risk studies require the use of analytical methods at the system level that considers subsystems and components in assessing their failure probabilities and consequences. Systematic, quantitative, qualitative or semi-quantitative approaches for assessing the failure probabilities and consequences of engineering systems are used for this purpose. A systematic approach allows an analyst to evaluate expediently and easily complex systems for safety and risk under different operational and extreme conditions. The ability to quantitatively evaluate these systems helps cut the cost of unnecessary and often expensive redesign, repair, strengthening or replacement of components, subsystems and systems. The results of risk analysis can also be utilized in decision analysis methods that are based on cost-benefit tradeoffs.

Risk assessment is a technical and scientific process by which the risks of a given situation for a system are modeled and quantified. Risk assessment can require and/or provide both qualitative and quantitative data to decision makers for use in risk management.

Risk assessment or risk analysis provides the process for identifying hazards, event-probability assessment, and consequence assessment. The risk assessment

Table 2 Risk perception

Activity or technology	League of women voters	College students	Experts
Nuclear power	1	1	20
Motor vehicles	2	5	1
Hand guns	3	2	4
Smoking	4	3	2
Motorcycles	5	6	6
Alcoholic beverages	6	7	3
General aviation	7	15	12
Police work	8	8	17
Pesticides	9	4	8
Surgery	10	11	5
Fire fighting	11	10	18
Large construction	12	14	13
Hunting	13	18	23
Spray cans	14	13	25
Mountain climbing	15	22	28
Bicycles	16	24	15
Commercial aviation	17	16	16
Electric (non-nuclear) power	18	19	9
Swimming	19	29	10
Contraceptives	20	9	11
Skiing	21	25	29
X-rays	22	17	7
High school or college sports	23	26	26
Railroads	24	23	19
Food preservatives	25	12	14
Food coloring	26	20	21
Power mowers	27	28	27
Prescription antibiotics	28	21	24
Home applications	29	27	22

process answers three basic questions: (1) What can go wrong? (2) What is the likelihood that it will go wrong? (3) What are the consequences if it does go wrong? Answering these questions requires the utilization of various risk methods as discussed in this chapter.

A risk assessment process should utilize experiences gathered from project personnel including managers, other similar projects and data sources, previous risk assessment models, experiences from other industries and experts, in conjunction with analysis and damage evaluation/prediction tools. A risk assessment process is commonly a part of a risk-based or risk-informed methodology that should be constructed as a synergistic combination of decision models, advanced probabilistic reliability analysis algorithms, failure consequence assessment methods, and conventional performance assessment methodologies that have been employed in related industry for performance evaluation and management. The methodology should realistically account for the various sources and types of uncertainty involved in the decision making process [4, 5].

In this section, a typical overall methodology is provided in the form of a workflow or block diagram. The various components of the methodology are described in subsequent sections. Figure 5 provides an overall description of a methodology for risk-based management of structural systems for the purpose of demonstration [7]. The methodology consists of the following primary steps:

- Definition of analysis objectives and systems;
- Hazard analysis, definition of failure scenarios, and hazardous sources and their terms;
- Collection of data in a lifecycle framework;
- Qualitative risk assessment;
- Quantitative risk assessment; and
- Management of system integrity through failure prevention and consequence mitigation using risk-based decision making.

In order to understand failure and the consequences of failure, the states of success need to be defined. With the development of the definition of success, one can begin to assess the likelihood of occurrence and causes of failures. Most of the information required to develop an estimate of the likelihood of failure might exist in maintenance and operating histories available on the systems and equipment, and based on judgment and expert opinion. This information might not be readily accessible, and its extraction from its current source might be difficult. Also, assembling it in a manner that is suitable for the risk-based methodology might be a challenge.

2.2 Risk Events and Scenarios

In order to adequately assess all risks associated with a project, the process of identification of risk events and scenarios is an important stage in risk assessment. Risk events and scenarios can be categorized as follows:

- *Technical, technological, quality, or performance risks*: Such as unproven or complex technology, unrealistic performance goals, and changes to the technology used or to the industry standards during the project.
- *Project-management risks*: Such as poor allocation of time and resources, inadequate quality of the project plan, and poor use of project-management disciplines.
- *Organizational risks*: Such as cost, time, and scope objectives that are internally inconsistent, lack of prioritization of projects, inadequacy or interruption of funding, resource conflicts with other projects in the organization, errors by individuals or by an organization, and inadequate expertise and experience by project personnel.

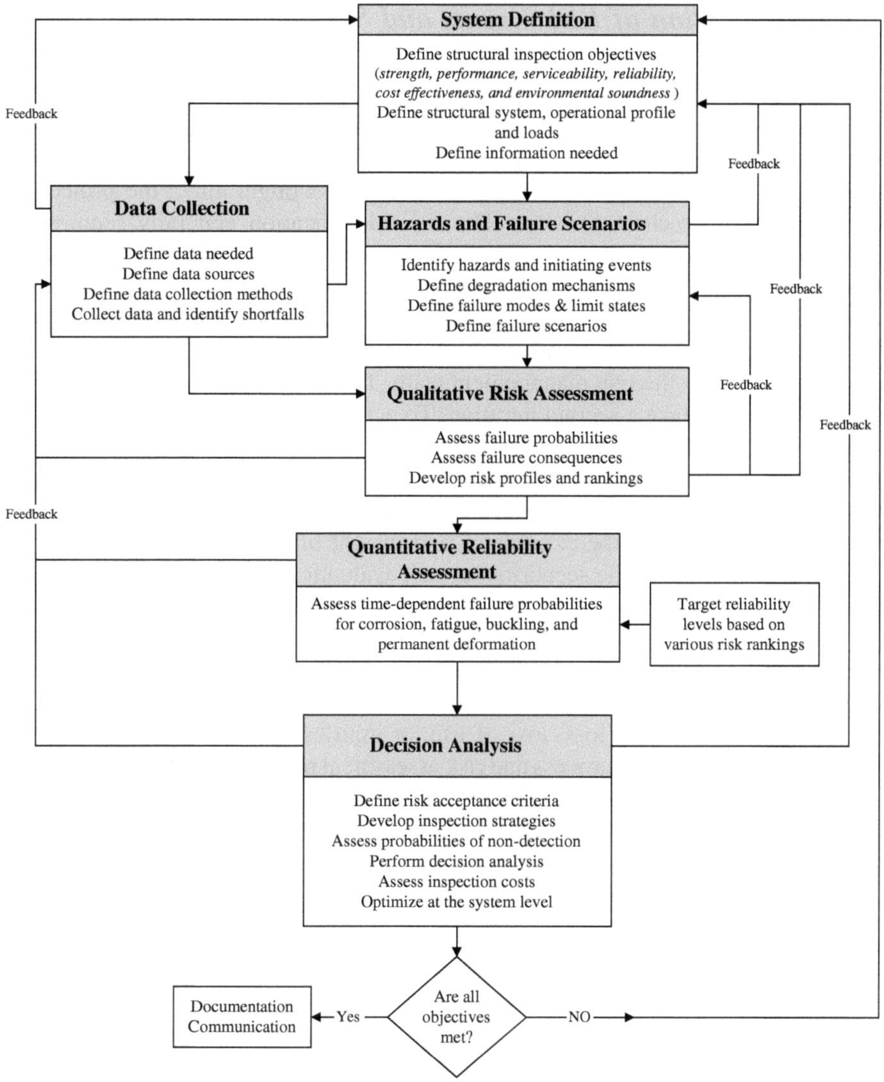

Fig. 5 Methodology for risk-based lifecycle management of structural systems

- *External risks*: Such as shifting legal or regulatory environment, labor issues, changing owner priorities, country risk, and weather.
- *Natural hazards*: Such as earthquakes, floods, strong wind, and waves generally require disaster recovery actions in addition to risk management.

Within these categories, several risk types can be identified.

2.3 Identification of Risk Events and Scenarios

The risk assessment process starts with the question "what can go wrong." The identification of what wrong can go entails defining hazards, risk events, and risk scenarios. The previous section provided categories of risk events and scenarios. Risk identification involves determining which risks might affect the project and documenting their characteristics. The risk identification generally requires the participation from a project team, risk management team, subject matter experts from other parts of the company, customers, end users, other project managers, stakeholders, and outside experts on as needed basis. Risk identification can be an iterative process. The first iteration may be performed by selected members of the project team, or by the risk management team. The entire project team and primary stakeholders may take a second iteration. To achieve an unbiased analysis, persons who are not involved in the project may perform the final iteration. Risk identification can be a difficult task, because it is often highly subjective, and there are no unerring procedures that may be used to identify risk events and scenarios other than relying heavily on the experience and insight of key project personnel.

The development of the scenarios for risk evaluation can be created deductively (e.g., fault tree) or inductively [e.g., failure mode and effect analysis (FMEA)] as provided in Table 3. The table shows methods of multiple uses including likelihood or frequency estimation expressed either deterministically or probabilistically. Also, they can be used to assess varying consequence categories including such items as: economic loss, loss of life, or injuries.

The risk identification process and risk assessment requires the utilization of these formal methods as shown in Table 3. These different methods contain similar approaches to answer the basic risk assessment questions; however, some techniques may be more appropriate than others for risk analysis depending on the situation.

2.4 Risk Breakdown Structure

Risk sources for a project can be organized and structured to provide a standard presentation that would facilitate understanding, communication and management. The previously presented methods can be viewed as simple linear lists of potential sources of risk, providing a set of headings under which risks can be arranged. These lists are sometimes called risk taxonomy. A simple list of risk sources might not provide the richness needed for some decision situations since it only presents a single level of organization. Some applications might require a full hierarchical approach to define the risk sources, with as many levels as are required to provide the necessary understanding of risk exposure. Defining risk sources in such a hierarchical structure is called a risk breakdown structure (RBS). The RBS is defined as a source-oriented grouping of project risks organized to define the total risk exposure of a project of interest. Each descending level represents an

Table 3 Risk assessment methods

Method	Scope
Safety/review audit	Identifies equipment conditions or operating procedures that could lead to a casualty or result in property damage or environmental impacts.
Checklist	Ensures that organizations are complying with standard practices.
What-if	Identifies hazards, hazardous situations, or specific accident events that could result in undesirable consequences.
Hazard and operability study (HAZOP)	Identifies system deviations and their causes that can lead to undesirable consequences and determine recommended actions to reduce the frequency and/or consequences of the deviations.
Preliminary hazard analysis (PrHA)	Identifies and prioritizes hazards leading to undesirable consequences early in the life of a system. It determines recommended actions to reduce the frequency and/or consequences of the prioritized hazards. This is an inductive modeling approach.
Probabilistic risk analysis (PRA)	Methodology for quantitative risk assessment developed by the nuclear engineering community for risk assessment. This comprehensive process may use a combination of risk assessment methods.
Failure modes and effects analysis (FMEA)	Identifies the components (equipment) failure modes and the impacts on the surrounding components and the system. This is an inductive modeling approach.
Fault tree analysis (FTA)	Identifies combinations of equipment failures and human errors that can result in an accident. This is a deductive modeling approach.
Event tree analysis (ETA)	Identifies various sequences of events, both failures and successes that can lead to an accident. This is an inductive modeling approach.
The Delphi technique	Assists to reach consensus of experts on a subject such as project risk while maintaining anonymity by soliciting ideas about the important project risks that are collected and circulated to the experts for further comment. Consensus on the main project risks may be reached in a few rounds of this process.
Interviewing	Identifies risk events by interviews of experienced project managers or subject matter experts. The interviewees identify risk events based on experience and project information.
Experience-based identification	Identifies risk events based on experience including implicit assumptions.
Brain storming	Identifies risk events using facilitated sessions with stakeholders, project team members, and infrastructure support staff.

increasingly detailed definition of risk sources for the project. The value of the RBS can be in aiding an analyst to understand the risks faced by the project.

An example RBS is provided in Table 4. In this example, four risk levels are defined as shown in the table. The project's risks are viewed as Level 0. Three types of Level 1 risks are provided in the table for the purpose of demonstration. The number of risk sources in each level varies and depends on the application at

Table 4 Risk breakdown structure for a project

Level 0	Level 1	Level 2	Level 3
Project Risks	Management	Corporate	History, experiences, culture, personnel
			Organization structure, stability, etc.
			Finances conditions
			Other projects
			⋮
		Customers and stakeholders	History, experiences, culture, personnel
			Contracts and agreements
			Requirement definition
			Finances and credit
			⋮
	External	Natural environment	Physical environment
			Facilities, site, equipment, materials
			Local services
			⋮
		Cultural	Political
			Legal, regulatory
			Interest groups
			Society and communities
			⋮
		Economic	Labor market, conditions, competition
			Financial markets
			⋮
	Technology	Requirements	Scope and objectives
			Conditions of use, users
			Complexity
			⋮
		Performance	Technology maturity
			Technology limitations
			New technologies
			New hazards or threats
			⋮
		Application	Organizational experience
			Personnel skill sets and experience
			Physical resources

hand. The subsequent Level 2 risks are provided in groups that are detailed further in Level 3. The RBS provides a means to systematically and completely identify all relevant risk sources for a project.

The risk breakdown structure should not be treated as a list of independent risk sources since commonly they have interrelations and common risk drivers. Identifying causes behind the risk sources is a key step towards an effective risk management plan including mitigation actions. A process of risk interrelation assessment and root-cause identification can be utilized to potentially lead to identifying credible scenarios that could lead to snowball effects for risk management purposes.

2.5 System Definition for Risk Assessment

Defining the system is an important first step in performing a risk assessment. A system can be defined as a deterministic entity comprising an interacting collection of discrete elements and commonly defined using deterministic models. The word "deterministic" implies that the system is identifiable and not uncertain in its architecture. The definition of the system is based on analyzing its functional and/or performance requirements. A description of a system may be a combination of functional and physical elements. Usually functional descriptions are used to identify high information levels on a system. A system may be divided into subsystems that interact. Additional detail leads to a description of the physical elements, components, and various aspects of the system.

The examination of a system needs to be made in a well-organized and repeatable fashion so that risk analysis can be consistently performed; therefore insuring that important elements of a system are defined and extraneous information is omitted. The formation of system boundaries is based upon the objectives of the risk analysis.

The establishment of system boundaries can assist in developing the system definition. The decision on what the system boundary is partially based on what aspects of the system's performance are of concern. The selection of items to include within the external boundary region is also reliant on the goal of the analysis. Beyond the established system boundary is the external environment of the system. Boundaries beyond the physical/functional system can also be established. For example, time may also be a boundary since an overall system model may change, as a product is further along in its lifecycle. The lifecycle of a system is important because some potential hazards can change throughout the lifecycle. For example, material failure due to corrosion or fatigue may not be a problem early in the life of a system; however, this may be an important concern later in the lifecycle of the system.

Along with identifying the boundaries, it is also important to establish a resolution limit for the system. The selected resolution is important since it limits the detail of the analysis. Providing too little detail might not provide enough information for the problem. Too much information may make the analysis more difficult and costly due to the added complexity. The depth of the system model needs to be sufficient for the specific problem. Resolution is also limited by the feasibility of determining the required information for the specific problem. For failure analysis, the resolution should be to the components level where failure data are available. Further resolution is not necessary and would only complicate the analysis.

The system breakdown structure is the top–down division of a system into subsystems and components. This architecture provides internal boundaries for the system. Often the systems/subsystems are identified as functional requirements that eventually lead to the component level of detail. The functional level of a system identifies the function(s) that must be performed for the operation of the system. Further decomposition of the system into "discrete elements" leads to the physical level of a system definition identifying the hardware within the system. By organizing a system hierarchy using a top–down approach rather than fragmentation of specific systems, a rational, repeatable and systematic approach to risk analysis can be achieved.

Further system analysis detail is addressed from modeling the system using some of the risk assessment methods described in Table 3. These techniques develop processes that can assist in decision making about the system. The logic of modeling based on the interaction of a system's components can be divided into induction and deduction. This difference in the technique of modeling and decision making is significant. Induction logic provides the reasoning of a general conclusion from individual cases. This logic is used when analyzing the effect of a fault or condition on a systems operation. Inductive analysis answers the question, "what are the system states due to some event?" In reliability and risk studies this "event" is some fault in the system. Several approaches using the inductive approach include: PrHA, FMEA, and ETA. Deductive approaches provide reasoning for a specific conclusion from general conditions. For system analysis this technique attempts to identify what modes of a system/subsystem/component failure can be used to contribute to the failure of the system. This technique answers the question, "how a system state can occur?" Inductive reasoning provides the techniques for FTA or its complement success tree analysis (STA).

2.6 Selected Risk Assessment Methods

2.6.1 Qualitative Versus Quantitative Risk Assessment

The risk assessment methods can be categorized according to how the risk is determined, by quantitative or qualitative analysis. Qualitative risk analysis uses judgment and sometimes "expert" opinion to evaluate the probability and consequence values. This subjective approach may be sufficient to assess the risk of a system, depending on the available resources.

Quantitative analysis relies on probabilistic and statistical methods, and databases that identify numerical probability values and consequence values for risk assessment. This objective approach examines the system in greater detail to assess risks.

The selection of a quantitative or qualitative method depends upon the availability of data for evaluating the hazard and the level of analysis needed to make a confident decision. Qualitative methods offer analyses without detailed information, but the intuitive and subjective processes may result in differences in outcomes by those who use them. Quantitative analysis generally provides a more uniform understanding among different individuals, but requires quality data for accurate results. A combination of both qualitative and quantitative analyses can be used depending on the situation.

Risk assessment requires estimates of the failure likelihood at some identified levels of decision making. The failure likelihood can be estimated in the form of lifetime failure likelihood, annual failure likelihood, mean time between failures, or failure rate. The estimates can be in numeric or non-numeric form. An example numeric form for an annual failure probability is 0.00015, and for a mean time between failures is 10 years. An example non-numeric form for "an annual failure likelihood" is large, and for a "mean time between failures" is medium. In the latter non-numeric form, guidance needs to be provided regarding the meaning of terms such as large, medium, small, very large, very small, etc. The selection of the form should be based on the availability of information, the ability of the personnel providing the needed information to express it in one form or another, and the importance of having numeric versus non-numeric information in formulating the final decisions.

The types of failure consequences that should be considered in a study need to be selected. They can include production loss, property damage, environmental damage, and safety loss in the form of human injury and death. Approximate estimates of failure consequences at the identified levels of decision making need to be determined. The estimates can be in numeric or non-numeric form. An example numeric form for production loss is 1,000 units. An example non-numeric form for production loss is large. In the latter non-numeric form, guidance needs to be provided regarding the meaning of terms such as large, medium, small, very large, very small, etc. The selection of the form should be based on the availability of information, the ability of the personnel providing the needed information to express it in one form or another, and the importance of having numeric versus non-numeric information in formulating the final decisions.

Risk estimates can be determined as a pair of the likelihood and consequences, and computed as the arithmetic multiplication of the respective failure likelihood and consequences for the equipment, components and details. Alternatively, for all cases, plots of failure likelihood versus consequences can be developed. Then, approximate ranking of them as groups according to risk estimates, failure likelihood, and/or failure consequences can be developed.

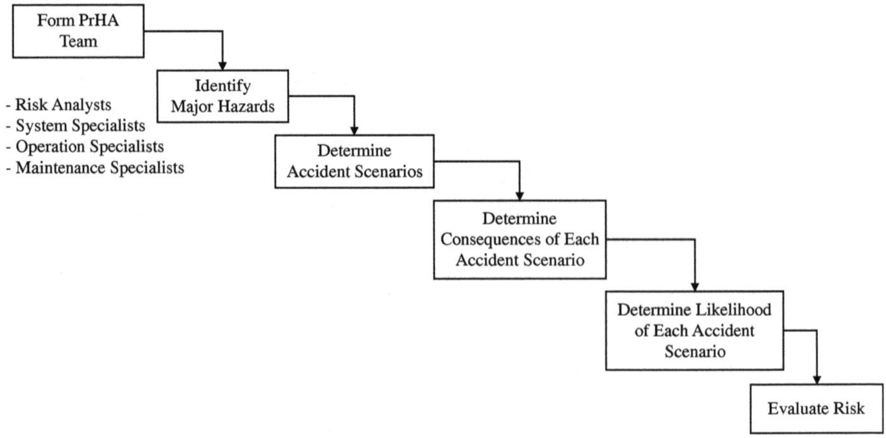

Fig. 6 Preliminary hazard analysis (PrHA) process

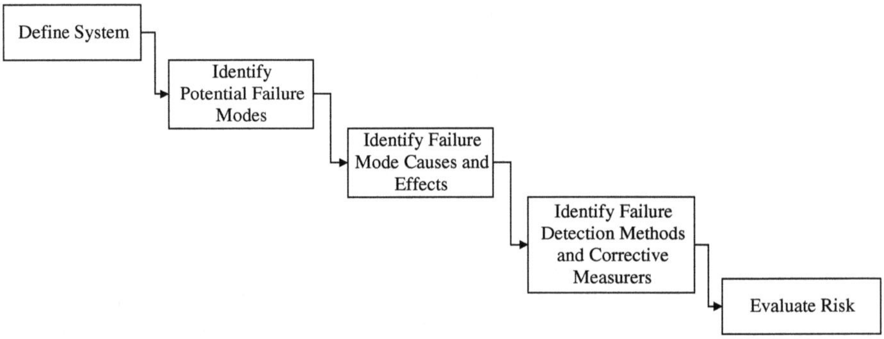

Fig. 7 Failure mode and effects analysis (FMEA) process

2.6.2 Preliminary Hazard Analysis

Preliminary Hazard Analysis (PrHA) is a common risk-based technology tool with many applications. The general process is shown in Fig. 6. This technique requires experts to identify and rank the possible accident scenarios that may occur. It is frequently used as a preliminary method to identify and reduce the risks associated with major hazards of a system.

2.6.3 Failure Mode and Effects Analysis

Failure Mode and Effects Analysis (FMEA) is another popular risk-based technology tool as shown in Fig. 7. This technique has been introduced both in the national and international regulations for the aerospace (US MIL-STD-1629A),

Table 5 Likelihood categories for a risk matrix

Category	Description	Annual probability range
A	Likely	$\geq$ 0.1 (1 in 10)
B	Unlikely	$\geq$ 0.01 (1 in 100) but < 0.1
C	Very unlikely	$\geq$ 0.001 (1 in 1,000) but < 0.01
D	Doubtful	$\geq$ 0.0001 (1 in 10,000) but < 0.001
E	Highly unlikely	$\geq$ 0.00001 (1 in 100,000) but < 0.0001
F	Extremely unlikely	< 0.00001 (1 in 100,000)

processing plant and marine industries. The Society of Automotive Engineers in its recommended practice introduces two types of FMEA: design and process FMEA. This analysis tool assumes a failure mode occurs in a system/component through some failure mechanism; the effect of this failure on other systems is then evaluated. A risk ranking can be developed for each failure mode for the effect on the overall performance of the system.

The various terms used in FMEA with examples based on the manufacturing of personal flotation devices (PFDs) are provided under subsequent headings to include failure mode, failure effect, severity rating, causes, occurrence rating, controls, detection rating and risk priority number.

2.6.4 Risk Matrices

Risk can be assessed and presented using matrices for preliminary screening by subjectively estimating probabilities and consequences in a qualitative manner. A risk matrix is a two-dimensional presentation of likelihood and consequences using qualitative metrics for both dimensions. According to this method, risk is characterized by categorizing probability and consequence on the two axes of a matrix. Risk matrices have been used extensively for screening of various risks. They may be used alone or as a first step in a quantitative analysis. Regardless of the approach used, risk analysis should be a dynamic process, i.e., a living process where risk assessments are re-examined and adjusted. Actions or inactions in one area can affect risk in another; therefore continuous updating is necessary.

The likelihood metric can be constructed using the categories shown in Table 5, whereas the consequences metric can be constructed using the categories shown in Table 6 with an example provided in Table 7. The consequence categories of Table 6 focus on the health and environmental aspects of consequences. The consequence categories of Table 7 focus on the economic impact, and should be adjusted to meet specific needs of industry and/or applications. An example risk matrix is shown in Fig. 8. In the figure, each boxed area is shaded depending on a subjectively assessed risk level. Three risk levels are used herein for illustration purposes of low (L), medium (M), and high (H). Other risk levels may be added using a scale of five levels instead of three levels if needed. These risk levels are

Table 6 Consequence categories for a risk matrix

Category	Description	Examples
I	Catastrophic	Large number of fatalities, and/or major long-term environmental impact.
II	Major	Fatalities, and/or major short-term environmental impact.
III	Serious	Serious injuries, and/or significant environmental impact.
IV	Significant	Minor injuries, and/or short-term environmental impact.
V	Minor	First aid injuries only, and/or minimal environmental impact.
VI	None	No significant consequence.

Table 7 Example consequence categories for a risk matrix in 2003 monetary amounts (US$)

Category	Description	Cost
I	Catastrophic loss	$\geq$ $10,000,000,000
II	Major loss	$\geq$ $1,000,000,000 but < $10,000,000,000
III	Serious loss	$\geq$ $100,000,000 but < $1,000,000,000
IV	Significant loss	$\geq$ $10,000,000 but < $100,000,000
V	Minor loss	$\geq$ $1,000,000 but < $10,000,000
VI	Insignificant loss	< $1,000,000

Probability Category							
	A	L	M	M	H	H	H
	B	L	L	M	M	H	H
	C						
	D	L	L	L	L	M	M
	E	L	L	L	L	L	M
	F						
		VI	V	IV	III	II	I
	Consequence Category						

Fig. 8 Example risk matrix

also called severity factors. The high (H) level can be considered as unacceptable risk level, the medium (M) level can be treated as either undesirable or as acceptable with review, and the low (L) level can be treated as acceptable without review.

2.6.5 Event Modeling: Event, Success Trees, and Fault Trees

Event modeling is a systematic, and often most complete, way to identify accident scenarios and quantify risk for risk assessment [1, 6, 8, 9]. This risk-based technology tool provides a framework for identifying scenarios to evaluate the performance of a system or component through system modeling. The combination of event tree analysis (ETA), success tree analysis (STA), and fault tree analysis (FTA) can provide a structured analysis to system safety.

Event tree analysis is often used if the successful operation of a component/system depends on a discrete (chronological) set of events. The initiating event is first followed by other events leading to an overall result (consequence). The ability to address a complete set of scenarios is developed since all combinations of both the success and failure of the main events are included in the analysis. The probability of occurrence of the main events of the event tree can be determined using a fault tree or its complement the success tree. The scope of the analysis for event trees and fault trees depends on the objective of the analysis.

Event tree analysis is appropriate if the operation of some system/component depends on a successive group of events. Event trees identify the various combinations of event successes and failures as a result of an initiating event to determine all possible scenarios. The event tree starts with an initiating event followed by some reactionary event. This reaction can either be a success or failure. If the event succeeds, the most commonly used indication is the upward movement of the path branch. A downward branch of the event tree marks the failure of an event. The remaining events are evaluated to determine the different possible scenarios. The scope of the events can be functions/systems that can provide some reduction to the possible hazards from the initiating event. The final outcome of a sequence of events identifies the overall state resulting from the scenario of events. Each path represents a failure scenario with varying levels of probability and risk. Different event trees can be created for different event initiators. Figure 9 shows an example event tree for the basic elements of a sprinkler system that might be critical for maintaining the integrity of a marine vessel.

Based on the occurrence of an initiating event, event tree analysis examines possible system outcomes or consequences. This analysis tool is particularly effective in showing interdependence of system components which is important in identifying events, that at first might appear insignificant, but due to the interdependency result in devastating results. Event tree analysis is similar to fault tree analysis because both methods use probabilistic reliability data of the individual components and events along each path to compute the likelihood of each outcome.

A quantitative evaluation of event tree probability values can be used for each event to evaluate the probability of the overall system state. Probability values for the success or failure of the events can be used to identify the probability for a specific event tree sequence. The probabilities of the events in a sequence can be provided as an input to the model or evaluated using fault trees. These probabilities for various events in a sequence can be viewed as conditional probabilities and therefore can be multiplied to obtain the occurrence probability of the sequence. The probabilities of various sequences can be summed up to determine the overall probability of a certain outcome. The addition of consequence evaluation of a scenario allows for generation of a risk value. For example, the occurrence probability of the top branch, i.e., scenario, in Fig. 9 is computed as the product of

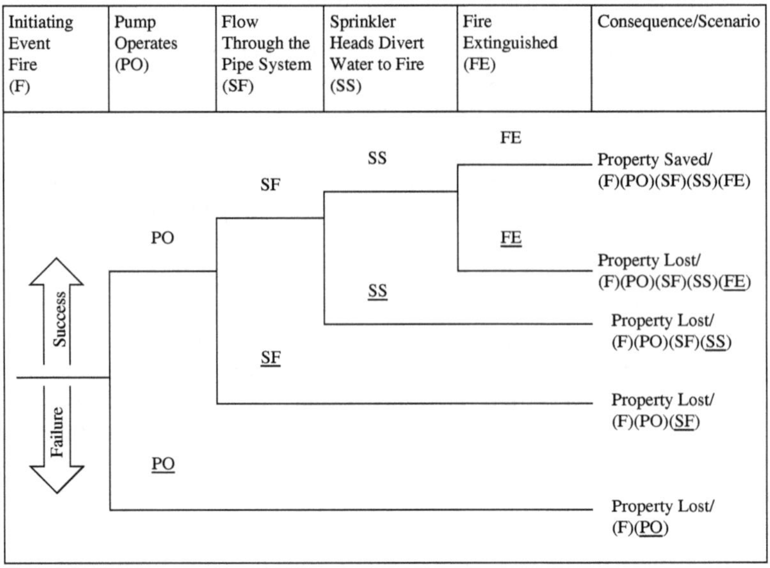

Fig. 9 Event tree example for sprinkler system

the probabilities of the composing events to this scenario, i.e., $F \cap PO \cap SF \cap SS \cap FE$ or (F)(PO)(SF)(SS)(FE) for short.

Complex systems are often difficult to visualize and the effect of individual components on the system as a whole is difficult to evaluate without an analytical tool. Two methods of modeling that have greatly improved the ease of assessing system reliability/risk are fault trees (FT) and success trees (ST). A fault tree is a graphical model created by deductive reasoning leading to various combinations of events that lead to the occurrence of some top event failure. A success tree shows the combinations of successful events leading to the success of the top event. A success tree can be produced as the complement (opposite) of the fault tree as illustrated in this section. Fault trees and success trees are used to further analyze the event tree headings (the main events in an event tree) to provide further detail to understand system complexities. In constructing the FT/ST only those failure/success events which are considered significant are modeled. This determination is assisted by defining system boundaries. For example, the event "pump operates (PO)" in Fig. 9 can be analyzed by developing a top–down logical breakdown of failure or success using fault tress or event trees, respectively.

Fault Tree Analysis (FTA) starts by defining a top event, which is commonly selected as an adverse event. An engineering system can have more than one top event. For example, a ship might have the following top events for the purpose of reliability assessment: power failure, stability failure, mobility failure, or structural failure. Then, each top event needs to be examined using the following logic: in order for the top event to occur, other events must occur. As a result, a set of lower

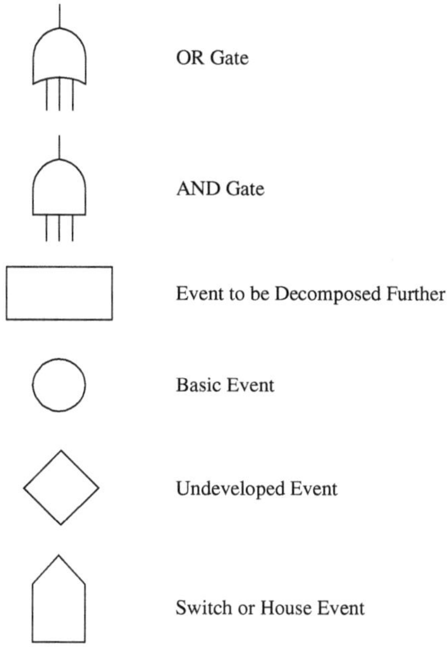

Fig. 10 Symbols used in fault tree analysis

level events is defined. Also, the form in which these lower level events are logically connected (i.e., in parallel or in series) needs to be defined. The connectivity of these events is expressed using "AND" or "OR" gates. Lower level events are classified into the following types:

- *Basic events*: These events cannot be decomposed further into lower level events. They are the lowest events that can be obtained. For these events, failure probabilities need be obtained.
- *Events that can be decomposed further*: These events can be decomposed further to lower levels. Therefore, they should be decomposed until the basic events are obtained.
- *Undeveloped events*: These events are not basic and can be decomposed further. However, because they are not important, they are not developed further. Usually, the probabilities of these events are very small or the effect of their occurrence on the system is negligible, or can be controlled or mediated.
- *Switch (or house) events*: These events are not random, and can be turned on or off with full control.

The symbols shown in Fig. 10 are used for these events. Also, a continuation symbol is shown, which is used to break up a fault tree into several parts for the purpose of fitting it in several pages.

FTA requires the development of a tree-looking diagram for the system that shows failure paths and scenarios that can result in the occurrence of a top event. The construction of the tree should be based on the building blocks and the Boolean logic gates.

The outcome of interest from the fault tree analysis is the occurrence probability of the top event. Since the top event was decomposed into basic events, its occurrence can be stated in the form of "AND," and "OR" of the basic events. The resulting statement can be restated by replacing the "AND" with the intersection of the corresponding basic events, and the "OR" with the union of the corresponding basic events. Then, the occurrence probability of the top event can be computed by evaluating the probabilities of the unions and intersections of the basic events. The dependence between these events also affects the resulting probability of the system.

For large fault trees, the computation of the occurrence probability of the top event can be difficult because of their size. In this case a more efficient approach is needed for assessing the reliability of a system, such as the minimal cut set approach. According to this approach, each cut set is defined as a set of basic events where the joint occurrence of these basic events results in the occurrence of the top event. A minimal cut set is a cut set with the condition that the non-occurrence of any one basic event from this set results in the non-occurrence of the top event. Therefore, a minimal cut set can be viewed as a subsystem in parallel. In general, systems have more than one minimal cut sets. The occurrence of the top event of the system can, therefore, be due to any one of these minimal cut sets. As a result, the system can be viewed as the union of all the minimal cut sets for the system. If probability values are assigned to the cut sets, a probability for the top event can be determined.

Common cause scenarios are events or conditions that result in the failure of seemingly separate systems or components. Common cause failures complicate the process of conducting risk analysis because a seemingly redundant system can be rendered ineffective by a common cause failure. For example, an emergency diesel generator fed by the same fuel supply as the main diesel engine will fail with the main diesel generator, if the fuel supply is the root source of the failure. The redundant emergency diesel generator is not truly redundant due to a common cause failure. Another example of common cause events is the failure of two separate but similar pieces of machinery due to a common maintenance problem, two identical pieces of equipment failing due to a common manufacturing defect, or two pieces of equipment failing due to a common environmental condition such as the flooding of a compartment or a fire in the vicinity of both pieces of machinery. A method for calculating the reliability of a system while taking into account common cause effects is the beta-factor model. Other methods include multiple Greek letter model, alpha factor model, and beta binomial failure rate model.

Part of risk-based decision analysis is pinpointing the system components that result in high-risk scenarios. Commercial system reliability software provides this type of analysis in the form of system reliability sensitivity factors to changes in the underlying component reliability values. In performing risk analysis, it is desirable to assess the importance of events in the model, or the sensitivity of final results to changes in the input failure probabilities for the events. Several sensitivity or importance factors are available and can be used. The most commonly used factors include (1) Fussell-Vesely factor, and (2) Birnbaum factor. Also, a weighted combination of these factors can be used as an overall measure.

2.7 Human Related Risks

Risk assessment requires the performance analysis of an entire system composed of a diverse group of components. The system definition readily includes the physical components of the system; however, humans are also part of most systems and provide significant contributions to risk. It has been estimated that nearly 90% of the accidents at sea are contributed to human error. The human contribution to risk can be estimated from an understanding of behavioral sciences. Both the "hardware failure" and human error should be addressed in the risk assessment since they both contribute to risks associated with the system. Once the human error probabilities are determined, human error/failures are treated in the same fashion as hardware failures in performing risk assessment quantification.

The determination of the human error contribution to risk is determined by human reliability analysis (HRA) tools. HRA is the discipline that enables the analysis and impact of humans on the reliability and safety of systems. Important results of HRA are determining the likelihood of human error as well as ways in which human errors can be reduced. When combined with system risk analysis, HRA methods provide an assessment of the detrimental effects of humans on the performance of the system. Human reliability analysis is generally considered to be composed of three basic steps: error identification, modeling, and quantification.

Other sources of human-related risks are in the form of deliberate sabotage of a system from within a system or as threat from outside the system, such as a computer hacker or a terrorist. The hazard in this case is not simply random but intelligent. The methods introduced in earlier sections might not be fully applicable for this risk type. The threat scenarios to the system in this case have a dynamic nature that are affected by the defense or risk mitigation and management scenarios that would be implemented by an analyst. The use of game theory methods might be needed in this case in combination with other risk analysis and management methods. Game theory is introduced in the last subsection herein.

2.7.1 Human Error Identification

Human errors are unwanted circumstances caused by humans that result in deviations from expected norms that place systems at risk. It is important to identify the relevant errors to make a complete and accurate risk assessment. Human error identification techniques should provide a comprehensive structure for determining significant human errors within a system. Quality HRA allows for accuracy in both the HRA assessment and overall system risk assessment.

Identification of human errors requires knowledge about the interactions of humans with other humans or machines (the physical world). It is the study of these interfaces that allows for the understanding of human errors. Potential sources of information for identifying human error may be determined from task analysis, expert judgment, laboratory studies, simulation and reports. Human errors may be considered active or latent depending on the time delay between when the error occurs and when the system fails.

It is important to note the distinction between human errors and human factors. Human errors are generally considered separately from human factors that applies information about human behavior, abilities, limitations, and other characteristics to the design of tools, machines, systems tasks, jobs, and environments for productive, safe, comfortable, and effective human use. Human factors are determined from performing descriptive studies for characterizing populations and experimental research. However, human factors analysis may contribute to the human reliability analysis.

2.7.2 Human Error Modeling

Once human errors have been identified they must be represented in a logical and quantifiable framework along with other components that contribute to the risk of the system. This framework can be determined from development of a risk model. Currently, there is no consensus on how to model human reliably. Many of these models utilize human event trees and fault trees to predict human reliability values. The identifications of human failure events can also be identified using Failure Mode and Effects Analysis. The human error rate estimates are often based on simulation tests, models, and expert estimation.

2.7.3 Human Error Quantification

Quantification of human error reliability promotes the inclusion of the human element in risk analysis. This is still a developing science requiring understanding of human performance, cognitive processing, and human perceptions. Since an exact model for human cognition has not been developed, much of the current human reliability data relies on accident databases, simulation and other empirical approaches. Many of the existing data sources were developed for from specific industry data such as nuclear and aviation industries. The application of these data sources for a specific problem should be thoroughly examined prior to application for a specific model. The result of the quantification of human reliability in terms of probability of occurrence is typically called a human error probability (HEP). There are many techniques that have been developed to help predict the HEP values. The Technique for Human Error Rate Prediction (THERP) is one of the most widely used methods for HEP. This technique is based on data gathered from the nuclear and chemical processing industries. THERP relies on HRA event tree modeling to identify the events of concern. Quantification is performed from data tables of basic HEP for specific tasks that may be modified based on the circumstances affecting performance.

The degree of human reliability is influenced by many factors often called performance shaping factors (PSF). PSFs are those factors that affect the ability of people to carry out required tasks. For example, the knowledge people have on how to don/activate a personal flotation device (PFD) will affect the performance of this task. Training (another PSF) in donning PFD's can also assist in the ability to perform this task. Another example is the training that is given to passengers on airplanes before takeoff on using seatbelts, emergency breathing devices, and flotation devices. Often the quantitative estimates of reliability are generated from a base error rate that is then altered based on the PSFs of the particular circumstances. Internal performance shaping factors are an individual's own attributes (experience, training, skills, abilities, attitudes) that affect the ability of the person to perform certain tasks. External PSFs are the dynamic aspects of situation, tasks, and system that affect the ability to perform certain tasks. Typical external factors include environmental stress factors (such as heat, cold, noise, situational stress, time of day), management, procedures, time limitations, and quality of person-machine interface. With these PSF it is easy to see the dynamic nature of HEP evaluation based on the circumstances of the analysis.

2.7.4 Reducing Human Errors

Error reduction is concerned with lowering the likelihood for error in an attempt to reduce risk. The reduction of human errors may be achieved by human factors interventions or by engineering means. Human factors interventions include improving training or improving the human–machine interface (such as alarms, codes, etc.) based on an understanding of the causes of error. Engineering means of error reduction may include automated safety systems or interlocks. The selection of the corrective actions to take can be done through decision analysis considering cost-benefit criteria.

2.7.5 Game Theory for Intelligent Threats

Game theory can be used to model human behavior, herein as a threat to a system. Generally game theory utilizes mathematics, economics, and the other social and behavioral sciences to model human behavior.

An example of intelligent threats is terrorism and sabotage as an ongoing battle between coordinated opponents representing a two-party game, where each opponent seeks to achieve their own objectives within a system. In the case of terrorism, it is a game of a well-established political system as a government versus an emerging organization that uses terrorism to achieve partial or complete dominance. Each player in this game seeks a utility, i.e., benefit, that is a function of the desired state of the system. In this case, maintaining system survival is the desired state for the government; whereas the opponent seeks a utility based on the failure state of the system. The government, as an opponent, is engaged in risk mitigation whose actions seek to reduce the threat, reduce the system vulnerability, and/or mitigate the consequences of any successful attacks. The terrorists, as an opponent, can be viewed as the aggressor who strives to alter or damage their opponent's desired system state. This game involves an intelligent threat, and is dynamic. The game is ongoing until the probability of a successful disruptive attempt of the aggressor reaches an acceptable level of risk; a stage where risk is considered under control, and the game is brought to an end. Classical game theory can be used in conjunction with probabilistic risk analysis to determine optimal mitigation actions that maximize benefits. In general, gaming could involve more than two players. The use of these concepts in risk analysis and mitigation needs further development and exploration.

2.8 Economic and Financial Risks

Economic and financial risks can be grouped into categories that include market risks, credit risks, operation risks, and reputation risks. Ayyub [3] provides additional information on these risks.

2.9 Data Needs for Risk Assessment

In risk assessment, the methods of probability theory are used to represent engineering uncertainties. In this context, it refers to event-occurrence likelihoods that occur with periodic frequency, such as weather, yet also to conditions which are existent but unknown, such as probability of an extreme wave. It applies to the magnitude of an engineering parameter, yet also to the structure of a model. By contrast, probability is a precise concept. It is a mathematical concept with an explicit definition. We use the mathematics of probability theory to represent uncertainties, despite that those uncertainties are of many forms.

The term probability has a precise mathematical definition, but its meaning when applied to the representation of uncertainties is subject to differing interpretations. The *frequentist* view holds that probability is the propensity of a physical system in a theoretically infinite number of repetitions; that is, the frequency of occurrence of an outcome in a long series of similar trials (e.g., the frequency of a coin landing heads-up in an infinite number of flips is the probability of that event). In contrast, the *Bayesian* view holds that probability is the rational degree of belief that one holds in the occurrence of an event or the truth of a proposition; probability is manifest in the willingness of an observer to take action upon this belief. This latter view of probability, which has gained wide acceptance in many engineering applications, permits the use of quantified professional judgment in the form of subjective probabilities. Mathematically, such subjective probabilities can be combined or operated on as any other probability.

Data are needed to perform quantitative risk assessment or provide information to support qualitative risk assessment. Information may be available if data have been maintained on a system and components of interest. The relevant information for risk assessment included the possible failures, failure probabilities, failure rates, failure modes, possible causes, and failure consequences. In the case of a new system, data may be used from similar systems if this information is available. Surveys are a common tool used to provide some means of data. Statistical analysis can be used to assess confidence intervals and uncertainties in estimated parameters of interest. Expert judgment may also be used as another source of data [2]. The uncertainty with the quality of the data should be identified to assist in the decision making process.

Data can be classified to including generic and project or plant specific types. Generic data are information from similar systems and components. This information may be the only information available in the initial stages of system design. Therefore, potential differences due to design or uncertainty may result from using generic data on a specific system. Plant specific data are specific to the system being analyzed. This information is often developed after the operation of a system. Relevant data need to be identified and collected as data collection can be costly. The data collected can then be used to update the risk assessment. Bayesian techniques can be used to combine objective and subjective data.

Data can be classified as failure probability data and failure consequence data. The failure probability data can include failure rates, hazard functions, times between failures, results from reliability studies, and any influencing factors and their effects. Failure-consequence data include loss reports, damages, litigation outcomes, repair costs, injuries, and human losses. Also, influencing factors, and effects of failure prevention and consequence mitigation plans. Areas of deficiency in terms of data availability should be identified, and sometimes failure databases need to be constructed. Data deficiency can be used as a basis for data collection and expert opinion elicitation.

3 Risk Management and Control

Adding risk control to risk assessment produces risk management. Risk management is the process by which system operators, managers, and owners make safety decisions, regulatory changes, and choose different system configurations based on the data generated in the risk assessment. Risk management involves using information from the previously described risk assessment stage to make educated decisions about system safety. Risk control includes failure prevention and consequence mitigation.

Risk management requires the optimal allocation of available resources in support of group goals. Therefore, it requires the definition of acceptable risk, and comparative evaluation of options and/or alternatives for decision making. The goals of risk management are to reduce risk to an acceptable level and/or prioritize resources based on comparative analysis. Risk reduction is accomplished by preventing an unfavorable scenario, reducing the frequency, and/or reducing the consequence. A graph showing the risk relationship should be treated as nonlinear curves. Moreover, the vertical axis is termed as probability whereas it is commonly expressed as an annual exceedence probability or frequency as shown in Fig. 1. In cases involving qualitative assessment, a matrix presentation can be used as shown in Fig. 8. The figure shows probability categories, severity categories, and risk ratings. A project's base value is commonly assumed as zero. Each risk rating value requires a different mitigation plan.

3.1 Risk Acceptance

Risk acceptance constitutes a definition of safety as discussed in previous sections. Therefore, risk acceptance is considered a complex subject that is often subject to controversial debate. The determination of acceptable levels of risk is important to determine the risk performance a system needs to achieve to be considered safe. If a system has a risk value above the risk acceptance level, actions should be taken to address safety concerns and improve the system through risk reduction measures. One difficulty with this process is defining acceptable safety levels for activities, industries, structures, etc. Since the acceptance of risk depends upon society perceptions, the acceptance criteria do not depend on the risk value alone. This section describes several methods that have been developed to assist in determining acceptable risk values as summarized in Table 8.

Risk managers make decisions based on risk assessment and other considerations including economical, political, environmental, legal, reliability, producibility, safety, and other factors. The answer to the question "How safe is safe enough?" is difficult and constantly changing due to different perceptions and understandings of risk. To determine "acceptable risk," managers need to analyze alternatives for the best choice. In some industries, an acceptable risk has been defined by consensus. For example, the U.S. Nuclear Regulatory Commission requires that reactors be designed such that the probability of a large radioactive release to the environment from a reactor incident shall be less than $1\text{x}10^{-6}$ per year. Risk levels for certain carcinogens and pollutants have also been given acceptable concentration levels based on some assessment of acceptable risk. However, risk acceptance for many other activities are not stated.

Often the level of risk acceptance with various activities is implied. Society has reacted to risks through the developed level of balance between risk and potential benefits. Measuring this balance of accepted safety levels for various risks provides a means for assessing society values. These threshold values of acceptable

Table 8 Methods for determining risk acceptance

Risk acceptance method	Summary
Risk conversion factors	This method addresses the attitudes of the public about risk through comparisons of risk categories. It also provides an estimate for converting risk acceptance values between different risk categories.
Farmers curve	It provides an estimated curve for cumulative probability risk profile for certain consequences (e.g., deaths). It demonstrates graphical regions of risk acceptance/non-acceptance.
Revealed preferences	Through comparisons of risk and benefit for different activities, this method categorizes society preferences for voluntary and involuntary exposure to risk.
Evaluation of magnitude of consequences	This technique compares the probability of risks to the consequence magnitude for different industries to determine acceptable risk levels based on consequence.
Risk effectiveness	It provides a ratio for the comparison of cost to the magnitude of risk reduction. Using cost-benefit decision criteria, a risk reduction effort should not be pursued if the costs outweigh the benefits. This may not coincide with society values about safety.
Risk comparison	The risk acceptance method provides a comparison between various activities, industries, etc., and is best suited to comparing risks of the same type.

risk depend on a variety of issues including the activity type, industry, and users, and the society as a whole.

Target risk or reliability levels are required for developing procedures and rules for ship structures. For example, the selected reliability levels determine the probability of failure of structural components. The following three methods were used to select target reliability values:

- Agreeing upon a reasonable value in cases of novel structures without prior history.
- Calibrating reliability levels implied in currently, successfully used design codes.
- Choosing target reliability level that minimizes total expected costs over the service life of the structure for dealing with design for which failure results in only economic losses and consequences.

The first approach can be based on expert opinion elicitation. The second approach called code calibration is the most commonly used approach as it provides the means to build on previous experiences. For example, rules provided by classification and industry societies can be used to determine the implied reliability and risk levels in respective rules and codes, then target risk levels can be set in a consistent manner, and new rules and codes can be developed to produce future designs and vessels that are of similar levels that offer reliability and/or risk consistency. The third approach can be based on economic and tradeoff analysis. In subsequent sections, the methods of Table 8 for determining risk acceptance are discussed.

Table 9 Risk conversion values for different risk factors

Risk factors	Risk conversion (RF) factor	Computed RF value
Origin	Natural/human-made	20
Severity	Ordinary/catastrophic	30
Volition	Voluntary/involuntary	100
Effect	Delayed/immediate	30
Controllability	Controlled/uncontrolled	5–10
Familiarity	Old/new	10
Necessity	Necessary/luxury	1
Costs	Monetary/non-monetary	NA
Origin	Industrial/regulatory	NA
Media	Low profile/high profile	NA

NA not available

3.1.1 Risk Conversion Factors

Analysis of risks shows that there are different taxonomies that demonstrate the different risk categories often called "risk factors." These categories can be used to analyze risks on a dichotomous scale comparing risks that invoke the same perceptions in society. For example, the severity category may be used to describe both ordinary and catastrophic events. Grouping events that could be classified as ordinary and comparing the distribution of risk to a similar grouping of catastrophic categories yields a ratio describing the degree of risk acceptance of ordinary events as compared to catastrophic events. The comparison of various categories determined the risk conversion values as provided in Table 9. These factors are useful in comparing the risk acceptance for different activities, industries, etc. By computing the acceptable risk in one activity, an estimate of acceptable risk in other activities can be calculated based on the risk conversion factors. A comparison of several common risks based on origin and volition is shown in Table 10.

3.1.2 Farmer's Curve

The Farmer's curve is graph of the cumulative probability versus consequence for some activity, industry or design as shown in Figs. 1 and 2. This curve introduces a probabilistic approach in determining acceptable safety limits. Probability (or frequency) and consequence values are calculated for each level of risk generating a curve that is unique to hazard of concern. The area to the right (outside) of the curve is generally considered unacceptable since the probability and consequence values are higher than the average value delineated by the curve. The area to the left (inside) of the curve is considered acceptable since probability and consequence values are less than the estimated value of the curve.

Table 10 Classification of common risks

		Voluntary	Involuntary		
Source	Size	Immediate	Delayed	Immediate	Delayed
Human Made	Catastrophic	Aviation		Dam failure	Pollution
				Building fire	Building fire
				Nuclear accident	
	Ordinary	Sports	Smoking	Homicide	
		Boating	Occupation		
		Automobiles	Carcinogens		
Natural	Catastrophic			Earthquakes	
				Hurricanes	
				Tornadoes	
				Epidemics	
	Ordinary			Lighting	Disease
				Animal bites	

3.1.3 Method of Revealed Preferences

The method of revealed preferences provides a comparison of risk versus benefit and categorization for different risk types. The basis for this relationship is that risks are not taken unless there is some form of benefit. Benefit may be monetary or some other item of worth such as pleasure. The different risk types are for the risk category of voluntary versus involuntary actions as shown in Fig. 11.

3.1.4 Magnitudes of Risk Consequence

Another factor affecting the acceptance of risk is the magnitude of consequence of the event that can result from some failure. In general, the larger the consequence is, the less the likelihood that this event may occur. This technique has been used in several industries to demonstrate the location of the industry within societies' risk acceptance levels based on consequence magnitude as shown in Fig. 12. Further evaluation has resulted in several estimates for the relationship between the accepted probability of failure and the magnitude of consequence for failure as provided by Allen in 1981 and called herein the CIRIA (Construction Industry Research and Information Association) equation:

$$P_f = 10^{-4}\frac{KT}{n} \tag{6}$$

where T is the life of the structure, K is a factor regarding the redundancy of the structure, and n is the number of people exposed to risk. Another estimate is Allen's equation that is given by:

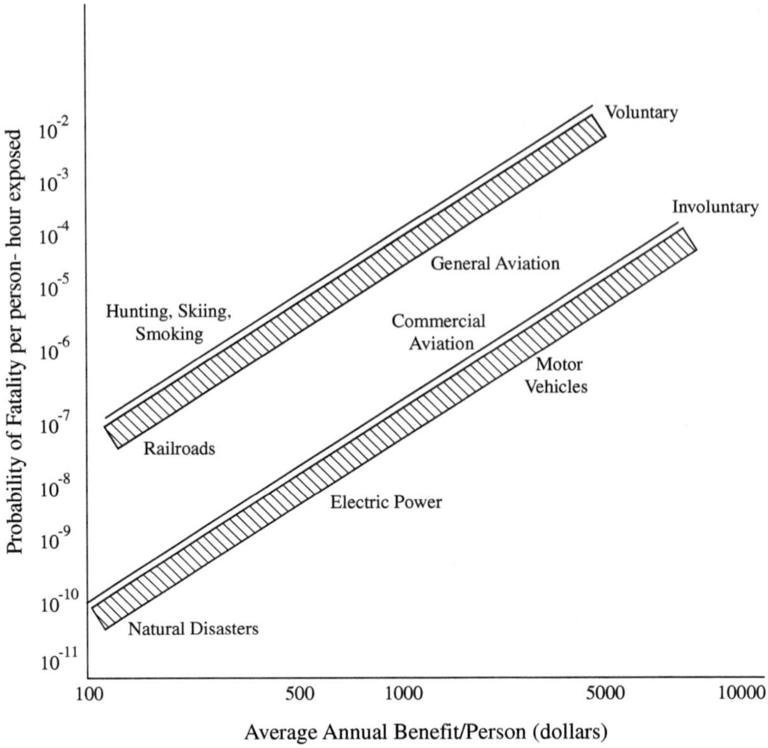

Fig. 11 Accepted risk of voluntary and involuntary activities

Fig. 12 Implicit risk-based on consequence of failure for industries

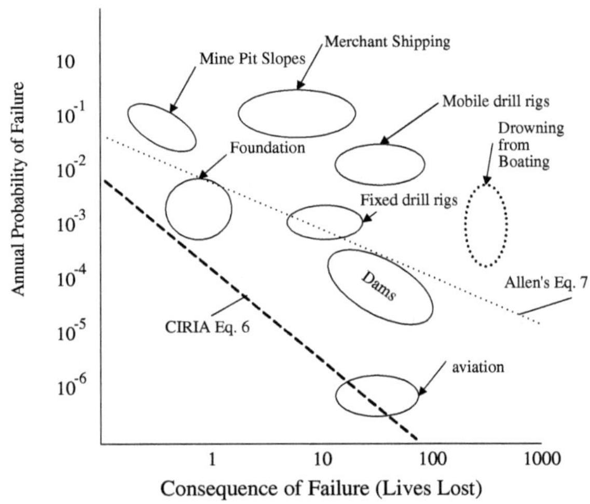

$$P_f = 10^{-5} \frac{TA}{W\sqrt{n}} \tag{7}$$

where T is the life of the structure, n is the number of persons exposed to risk, and A and W are factors regarding the type and redundancy of the structure. Equation 6 offers a lower bound, whereas Eq. 7 offers a middle lime.

3.1.5 Risk Reduction Cost Effectiveness Ratio

Another measuring tool to assess risk acceptance is the determination of risk reduction effectiveness:

$$Risk\ \mathrm{R}eduction\ Effectiveness = \frac{Cost}{\Delta\ Risk} \tag{8}$$

where the cost should be attributed to risk reduction, and $\Delta Risk$ is the level of risk reduction as follows:

$$\Delta\ Risk = (\text{Risk before mitigation action}) - (\text{Risk after mitigation action}) \tag{9}$$

The difference in Eq. 9 is also called the benefit attributed to a risk reduction action. Risk effectiveness can be used to compare several risk reduction efforts. The initiative with the smallest risk effectiveness provides the most benefit for the cost. Therefore, this measurement may be used to help determine an acceptable level of risk. The inverse of this relationship may also be expressed as cost effectiveness.

3.1.6 Risk Comparisons

This technique uses the frequency of severe incidents to directly compare risks between various areas of interest to assist in justifying risk acceptance. Risks can be presented in different ways that can impact how the data are used for decisions. Often values of risk are manipulated in different forms for comparison reasons as demonstrated in Table 11. Comparison of risk values should be taken in the context of the values' origin and uncertainties involved.

This technique is most effective for comparing risks that invoke the same human perceptions and consequence categories. Comparing risks of different categories is cautioned since the differences between risk and perceived safety may not provide an objective analysis of risk acceptance. The use of risk conversion factors may assist in transforming different risk categories. Conservative guidelines for determining risk acceptance criteria can be established for voluntary risks to the public from the involuntary risk of natural causes.

3.2 Decision Analysis

Decision analysis provides a means for systematically dealing with complex problems to arrive at a decision. Information is gathered in a structured manner to provide the best answer to the problem. A decision generally deals with three elements: alternatives, consequences, and preferences. The alternatives are the possible choices for consideration. The consequences are the potential outcomes of a decision. Decision analysis provides methods for quantifying preference trade-offs for performance along multiple decision attributes while taking into account risk objectives. Decision attributes are the performance scales that measure the degree to which objectives are satisfied. For example, one possible attribute is reducing lives lost for the objective of increasing safety. Additional examples of objectives may include minimize the cost, maximize utility, maximize reliability, and maximize profit. The decision outcomes may be affected by uncertainty; however, the goal is to choose the best alternative with the proper consideration of uncertainty. The analytical depth and rigor for decision analysis depends on the desired detail in making the decision. Cost-benefit analysis, decision trees, influence diagrams and the analytic hierarchy process are some of the tools to assist in decision analysis. Also, decision analysis should consider constraints, such as availability of system for inspection, availability of inspectors, preference of certain inspectors, and availability of inspection equipment.

3.3 Benefit-Cost Analysis

Risk managers commonly weigh various factors including cost and risk. Risk-benefit analysis can also be used for risk management. Economic efficiency is important to determine the most effective means of expending resources. At some point the costs for risk reduction do not provide adequate benefit. This process compares the costs and risk to determine where the optimal risk value is on a cost basis. This optimal value occurs, as shown in Fig. 13, when costs to control risk are equal to the risk cost due to the consequence (loss). Investing resources to reduce low risks below this equilibrium point is not providing a financial benefit. This technique may be used when cost values can be attributed to risks. This analysis might be difficult to perform for certain risk such as risk to human health and environmental risks since the monetary values are difficult to estimate for human life and the environment.

The present value of incremental costs and benefits can be assessed and compared among alternatives that are available for risk mitigation or system design. Several methods are available to determine which, if any, option is most worth pursuing. In some cases, no alternative will generate a net benefit relative to the base case. Such a finding would be used to argue for pursuit of the base case scenario. The following are the most widely used present value comparison

Table 11 Ways to identify risk of death

Ways to identify risk of death	Summary
Number of fatalities	This measure shows the impact in terms of the number of fatalities on society. Comparison of these values is cautioned since the number of persons exposed to the particular risk may vary. Also, the time spent performing the activity may vary. Different risk category types should also be considered to compare fatality rates.
Annual mortality rate/ individual	This measure shows the mortality risk normalized by the exposed population. This measure adds additional information about the number of exposed persons; however, the measure does not include the time spent on the activity.
Annual mortality	This measure provides the most complete risk value since the risk is normalized by the exposed population and the duration of the exposure.
Loss of life exposure (LLE)	This measure converts a risk into a reduction in the expected life of an individual. It provides a good means of communicating risks beyond probability values.
Odds	This measure is a layman format for communicating probability, for example, 1 in 4.

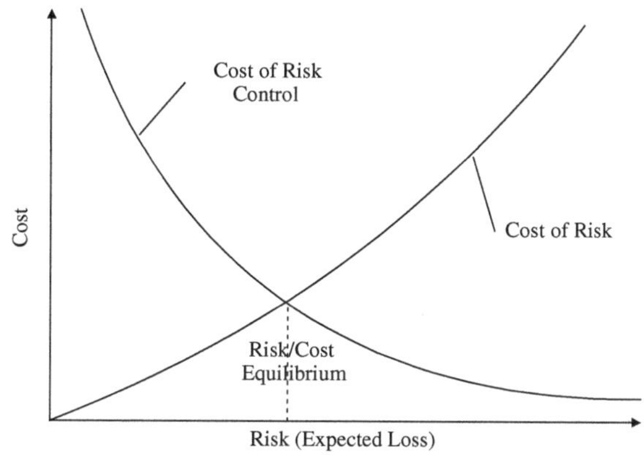

Fig. 13 Comparison of risk and control costs

methods: (1) net present value (NPV), (2) benefit-cost ratio, (3) internal rate of return, and (4) payback period. The net present value (NPV) method requires that each alternative need to meet the following criteria to warrant investment of funds: (1) having a positive NPV; and (2) having the highest NPV of all alternatives considered. The first condition insures that the alternative is worth undertaking relative to the base case, e.g., it contributes more in incremental benefits than it absorbs in incremental costs. The second condition insures that maximum benefits are obtained in a situation of unrestricted access to capital funds. The NPV can be calculated as follows:

$$NPV = \sum_{t=0}^{k} \frac{(B-C)_t}{(1+r)^t} = \sum_{t=0}^{k} \frac{B_t}{(1+r)^t} - \sum_{t=0}^{k} \frac{C_t}{(1+r)^t} \tag{10}$$

where B is future annual benefits in constant dollars, C is future annual costs inconstant dollars, r is annual real discount rate, k is number of years from the base year over which the project will be evaluated, and t is an index running from 0 to k representing the year under consideration.

The benefit of a risk mitigation action can be assessed as follows:

$$\text{Benefit} = \text{unmitigated risk} - \text{mitigated risk} \tag{11}$$

The cost in Eq. 10 is the cost of the mitigation action. The benefit minus the cost of mitigation can be used to justify the allocation of resources. The benefit-to-cost ratio can be computed, and may also be helpful in decision making. The benefit-to-cost ratio can be computed as

$$\text{Benefit-to-Cost Ratio } (B/C) = \frac{\text{Benefit}}{\text{Cost}} = \frac{\text{Unmitigated Risk} - \text{Mitigated Risk}}{\text{Cost of Mitigation Action}} \tag{12}$$

Ratios greater than one are desirable. In general, the larger the ratio, the better the mitigation action.

Accounting for the time value of money would require defining the benefit-cost ratio as the present value of benefits divided by the present value of costs. The benefit-cost ratio can be calculated as follows:

$$B/C = \frac{\sum_{t=0}^{k} \frac{B_t}{(1+r)^t}}{\sum_{t=0}^{k} \frac{C_t}{(1+r)^t}} \tag{13}$$

where B_t is future annual benefits in constant dollars, C_t is future annual costs inconstant dollars, r is annual real discount rate, and t is an index running from 0 to k representing the year under consideration. A proposed activity with a *B*/*C* ratio of discounted benefits to costs of one or more is expected to return at least as much in benefits as it costs to undertake, indicating that the activity is worth undertaking.

The internal rate of return (*IRR*) is defined as the discount rate that makes the present value of the stream of expected benefits in excess of expected costs zero. In other words, it is the highest discount rate at which the project will not have a negative *NPV*. To apply the *IRR* criterion, it is necessary to compute the *IRR* and then compare it with a base rate of, say, a 7% discount rate. If the real *IRR* is less than 7%, the project would be worth undertaking relative to the base case. The *IRR* method is effective in deciding whether or not a project is superior to the base case; however, it is difficult to utilize it for ranking projects and deciding among mutually exclusive alternatives. Project rankings established by the *IRR* method might be inconsistent with those of the *NPV* criterion. Moreover, a project might have more than one *IRR* value, particularly when a project entails major final costs, such as clean-up costs. Solutions to these limitations exist in capital budgeting procedures and practices that are often complicated or difficult to employ in practice and present opportunities for error.

The payback period measures the number of years required for net undiscounted benefits to recover the initial investment in a project. This evaluation method favors projects with near-term and more certain benefits, and fails to consider benefits beyond the payback period. The method does not provide information on whether an investment is worth undertaking in the first place.

The previous models for benefit-cost analysis presented in this section do not account for the full probabilistic characteristics of B and C in their treatment. Concepts from reliability assessment four can be used for this purpose. Assuming B and C to normally distributed, a benefit-cost index ($\beta_{B/C}$) can be defined as follows:

$$\beta_{B/C} = \frac{\mu_B - \mu_C}{\sqrt{\sigma_B^2 + \sigma_C^2}} \tag{14}$$

where μ and σ are the mean and standard deviation. The failure probability can be computed as

$$P_{f,B/C} = P(C > B) = 1 - \Phi(\beta) \tag{15}$$

In the case of lognormally distributed B and C, the benefit-cost index ($\beta_{B/C}$) can be computed as

$$\beta_{B/C} = \frac{\ln\left(\frac{\mu_B}{\mu_C}\sqrt{\frac{\delta_C^2+1}{\delta_B^2+1}}\right)}{\sqrt{\ln[(\delta_B^2 + 1)(\delta_C^2 + 1)]}} \tag{16}$$

where δ is the coefficient of variation. Equation 16 also holds for the case of lognormally distributed B and C. In the case of mixed distributions or cases involving basic random variables of B and C, the advanced second moment method or simulation method can be used. In cases where benefit is computed as revenue minus cost, benefit might be correlated with cost requiring the use of other methods.

3.4 Risk Mitigation

A risk mitigation strategy can be presented from a financial point of view. Risk mitigation in this context can be defined as an action to either reduce the probability of an adverse event occurring or to reduce the adverse consequences if it does occur. This definition captures the essence of an effective management process of risk. If implemented correctly a successful risk mitigation strategy should reduce any adverse (or downside) variations in the financial returns from a project, which are usually measured by either (1) the net present value (NPV) defined as the difference between the present value of the cash flows generated by a project and its capital cost and calculated as part of the process of assessing and appraising investments, or (2) the internal rate of return (IRR) defined as the return that can be earned on the capital invested in the project, i.e., the discount rate that gives an NPV of zero, in the form of the rate that is equivalent to the yield on the investment.

Risk mitigation involves direct costs like increased capital expenditure or the payment of insurance premiums; hence might reduce the average overall financial returns from a project. This reduction is often a perfectly acceptable outcome, given the risk aversion of many investors and lenders. A risk mitigation strategy is the replacement of an uncertain and volatile future with one where there is less exposure to adverse risks and so less variability in the return, although the expected NPV or IRR may be reduced. These two aspects are not necessarily mutually exclusive. Increasing risk efficiency by simultaneously improving the expected NPV or IRR and simultaneously reducing the adverse volatility is sometimes possible and should be sought. Risk mitigation should cover all phases of a project from inception to closedown or disposal.

Four primary ways are available to deal with risk within the context of a risk management strategy as follows: (1) risk reduction or elimination, (2) risk transfer, e.g., to a contractor or an insurance company, (3) risk avoidance, and (4) risk absorbance or pooling. Risk reduction or elimination is often the most fruitful form for exploration. A general principle of an effective risk management strategy is that commercial risks in projects and other business ventures should be borne wherever possible by the party that is best able to manage them, and thus mitigate the risks. Contracts and financial agreements are the principal forms to transfer risks. A most intuitive way of avoiding a risk is to avoid undertaking the project in a way that involves that risk. Cases where risks cannot, or cannot economically, be eliminated, transferred or avoided, they must be absorbed if the project is to proceed. Risk can be mitigated through proper uncertainty characterization. The presence of improperly characterized uncertainty could lead to higher adverse event-occurrence likelihood and consequences. Also, it could result in increasing estimated cost margins as a means of compensation.

4 Risk Communication

Risk communication can be defined as an interactive process of exchange of information and opinion among stakeholders such as individuals, groups, and institutions. It often involves multiple messages about the nature of risk or expressing concerns, opinions, or reactions to risk managers or to legal and institutional arrangements for risk management. Risk communication greatly affects risk acceptance and defines the acceptance criteria for safety.

The process of risk communication can be enhanced and improved in three aspects: (1) the process, (2) the message, and (3) the consumers. The risk assessment and management process needs to have clear goals with openness, balance, and competence. The contents of the message should account for audience orientation and uncertainty, provide risk comparison, and be complete. There is a need for consumer's guides that introduce risks associated with a specific technology, the process of risk assessment and management, acceptable risk, decision making, uncertainty, costs and benefits, and feedback mechanisms. Improving risk literacy of consumers is an essential component of the risk communication process.

The USACE has a 1992 Engineering Pamphlet (EP) on risk communication (EP 1110-2-8). The following are guiding considerations in communicating risk:

- Risk communication must be free of jargon,
- Consensus of expert needs to be established,
- Materials cited, and their sources must be credible,
- Materials must be tailored to audience,
- The information must be personalized to the extent possible,
- Motivation discussion should stress a positive approach and the likelihood of success, and
- Risk data must be presented in a meaningful manner.

References

1. Ang AH-S, Tang WH (1990) Probability concepts in engineering planning and design: volume 2 - decision risk and reliability. Wiley, New York
2. Ayyub BM (2002) Elicitation of expert opinions for uncertainty and risks. CRC Press, Boca Raton
3. Ayyub BM (2003) Risk analysis in engineering and economics. Chapman and Hall/CRC Press, Boca Raton
4. Ayyub BM, Klir GJ (2006) Uncertainty modeling and analysis for engineers and scientists. Chapman and Hall/CRC Press, Boca Raton
5. Ayyub BM, McCuen R (2003) Probability, statistics and reliability for engineers and scientists. Chapman and Hall/CRC Press, Boca Raton
6. Kumamoto H, Henley EJ (1996) Probabilistic risk assessment and management for engineers and scientists. IEEE Press, New York
7. Maryland Emergency Management Agency (MEMA) (2006) State of Maryland guide for the protection of critical infrastructure and key resources for homeland security, volume 1: critical asset and portfolio risk assessment (CAPRA) methodology. Office of Homeland Security, Annapolis
8. Modarres M (1993) What every engineer should know about reliability and analysis. Marcel Dekker Inc, New York
9. Modarres M, Kaminskiy M, Krivstov V (1999) Reliability engineering and risk analysis: a practical guide. Marcel Decker Inc., New York

Reliability Analysis of Gas Turbine

Fernando Jesus Guevara Carazas and Gilberto Francisco Martha de Souza

Abstract This chapter presents the application of reliability concepts to evaluate the overall performance of gas turbine used in open cycle or combined-cycle thermal power plants. The thermodynamics derived performance parameters of gas turbines are presented, including the presentation of the tests codes used to evaluate turbine performance during power plant commissioning. The IEEE Std 762 is presented as traditional way to measure gas turbine availability although it is based on deterministic analysis and does not allow long-term prediction of gas turbine performance. The fundamental reliability and availability concepts associated with gas turbine are presented and an example of reliability analysis of a heavy duty gas turbine is also presented.

1 Gas Turbine

The accelerated evolution of gas turbines from the 1940s was basically in the interest of the armed forces that used that piece of equipment as propulsion engines for combat aircraft during World War II. Gas turbine engines quickly replaced the reciprocating engines use at that time. After demostrating the effectiveness of gas turbine as propulsion engines for military aircraft, that equipment began to be installed on commercial aircraft designed from the 1950s. In this way, later in 1960 and 1970, most commercial aircraft, medium and wide bodies, such as the Boeing 707, McDonnell Douglas DC-8, Airbus A300, Boeing 747,

F. J. G. Carazas · G. F. M. de Souza (✉)
Department of Mechatronics and Mechanical Systems Engineering, Polytechnic School, University of São Paulo, Av. Prof. Mello Moraes 2231, 05508-900 São Paulo, Brazil
e-mail: gfmsouza@usp.br

G. F. M. de Souza (ed.), *Thermal Power Plant Performance Analysis*,
Springer Series in Reliability Engineering, DOI: 10.1007/978-1-4471-2309-5_8,

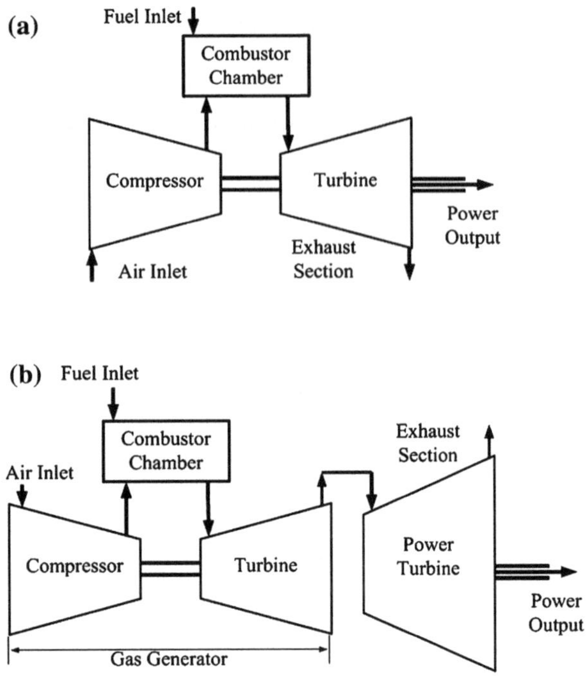

Fig. 1 Gas turbine configurations **a** Simple gas turbine system **b** Aero-derivative gas turbine configuration

McDonnell Douglas DC-10 among others, used gas turbine as propulsion engines. In parallel to the evolution of aircraft industry, the concepts of reliability applied to design and maintenance of those aircraft had occupied an important place. Thus, at the request of the Federal Aviation Agency (FAA) in 1965 was initiated a work to re-evaluate maintenance procedures of the Boeing 747, which later would give rise to the current philosophy of reliability-centred maintenance (RCM) [18, 22].

Gas turbines are complex systems of energy transformation. Just like any internal combustion engine they are composed of hundreds of components. However all gas turbines have three main sections that carry out the processes needed for the transformation of the fuel's chemical energy into mechanical energy. First, the compression section, where atmospheric air incomes and, by the compression process, the air pressure and temperature are increased. Next, the compressed air is driven to a combustion section where fuel is injected and burnt, increasing the temperature at constant pressure. Finally, the combustion products at high temperature are expanded in the turbine blades section generating net shaft power. In the case that the gas turbine is used for electric power generation, the shaft power is used to drive an electrical generator. The combustion products are exhausted through a nozzle into the atmosphere. The basic principle of gas turbine operation is show in Fig. 1.

2 Gas Turbine Applications

Besides to the aeronautic industry, the technology of gas turbine was gaining application in other areas of industry. In 1950 the Westinghouse Company began to produce gas turbines for industrial use. These turbines, despite using the same operating principle of aeronautical gas turbines, present some differences to those used in the aviation industry. The gas turbines are classified into two main groups: Aero-derivative and Heavy-Duty Gas Turbine.

2.1 Aero-Derivative Gas turbine (Aircraft-Derivative)

These turbines are derived from aeronautical designs and adapted for industrial work, consisting of two basic components: an aircraft-derivative gas generator, and a free-power turbine [7]. This configuration is show in Fig. 1b.

Once aero-derivative gas turbines derived from aircraft designs they are considered more reliable and economic, taking in view that it is easy to adapt existing technology to develop new technology for large gas turbines. Aero-derivative turbines currently reach around 40 MW. The main applications of such gas turbines are in the petrochemical industry and power generation industry, for combined-cycle power generation and cogeneration plants. In the petroleum industry, mainly in offshore platform, they are used for power generation and as prime movers for compressors and pumps. Finally, the aero-derivative gas turbines are used in military ship propulsion.

2.2 Frame Type Heavy-Duty Gas Turbine

These gas turbines began to be produced after the World War II. Nowadays the power output varies from 20 MW to around 250 MW with efficiencies around 40%, according to manufacturers such as GE and Siemens. This type of turbine has single shaft configuration, so that part of the energy produced by the turbine is transmitted to the compressor and the remaining energy is used as power output on the shaft. This configuration meets the needs of high load and constant speed typical from electric power generation [16]. This configuration is shown in Fig. 1a.

The industrial heavy-duty gas turbines employ compressors with 15–18 stages of axial-flow compression. Combustion chambers often use ring-shaped frame installed around the turbine. These turbines are characterized by high fuel flexibility, including dual fuel operation. Another feature of modern gas turbines is their low emission of NO_x. Manufacturers development programs have focused on evolutionary combustion systems capable of meeting the extremely low NO_x

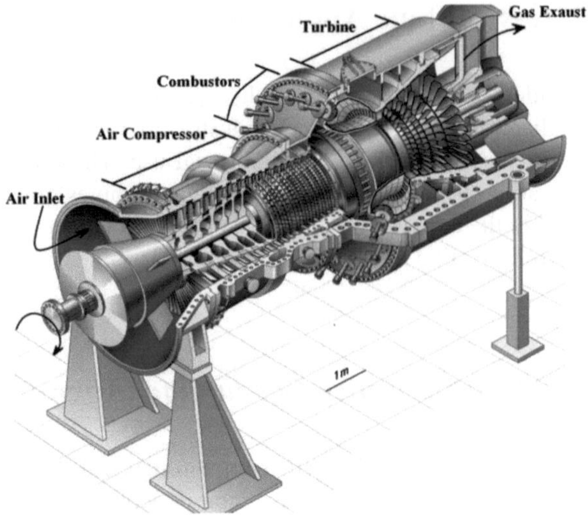

Fig. 2 Heavy-duty gas turbine (Adapted from Siemens, 2006)

levels required to meet current and future environmental regulations. Heavy-duty gas turbines are mainly used for pure power generation, including combined-cycle, and industrial cogeneration [2]. In Fig. 2 it is shown a heavy-duty gas turbine.

For power generation gas turbines are frequently used in both open cycle and combined-cycle configurations. Typically the open cycle gas turbine (gas turbine is operated alone) is used for reserve or peak generating capacity and is operated for a limited number of hours per year, between 2,000 and 5,000 h [5]. Usually the open cycle gas turbine power plants are converted on the combined-cycle configuration over the years. This change must be planned from the beginning of plant design, particularly in relation to the space needed for the installation of combined-cycle equipment. The simple cycle configuration is show in Fig. 3a.

The combined-cycle power plant consists of one or more gas turbines linked to an electrical generator and the exhaust gas flow into one or more heat recovery steam generator (HRSG). The exhaust gas from the gas turbine is directed through an HRSG that generates steam at one or more pressure levels (nowadays up to 3 pressure levels). The steam is fed to a steam turbine that drives a dedicated electrical generator [5].

For operation in combined cycle the exhaust gas temperature of heavy-duty gas turbines is usually higher than that of aero-derivative gas turbines. Additionally the exhaust flow in the heavy-duty gas turbines is high, allowing more steam generation in HRSG at higher temperatures. For this reason it is possible to generate more electrical energy with steam turbine. The combined-cycle configuration is show in Fig. 3b.

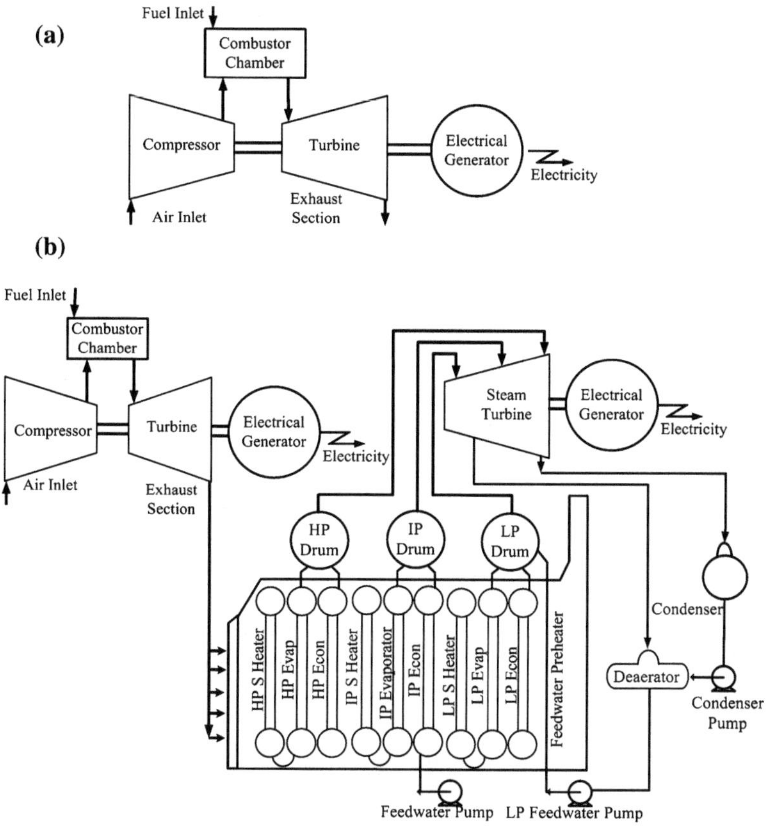

Fig. 3 Gas turbines thermoelectric driven power plants configurations **a** Simple cycle power plant configuration **b** Combined cycle power plant configuration

3 Performance, Reliability and Availability of Gas Turbines

Considering that gas turbines are derived from the aerospace industry, they are designed to operate with high reliability and high performance. However an industrial turbine must present an important feature specially when used for power generation, which is availability. That index represents the percentage of given period of time, expressed in hours, that the unit is in service and capable of providing a given nominal output.

Table 1 Gas turbine performance parameters

	Performance Parameter	Component	Composition	Unit
1	Compressor inlet temperature	Compressor	T_2	°F or °C
2	Compression Rate	Compressor	P_3/P_2	
3	Turbine inlet temperature	Compressor	T_4	°F or °C
4	Exhaust temperature	Turbine	$T_5 = T_6$	°F or °C
5	Exhaust gas flow	Turbine	$\dot{m}_5{}^a$	lb/h
6	Exhaust heat	Turbine	$\dot{m}_5 \times T_5$	Mbtu/h
7	Inlet pressure loss	Compressor	$P_1 - P_2$	In. H_2O
8	Exhaust pressure loss	Turbine	$P_5 - P_6$	In. H_2O

[a] $\dot{m}_5 = (\dot{m}_1 + \dot{m}_F + \dot{m}_W - \dot{m}_L)$

3.1 Gas Turbine Performance Parameters

Typically the gas turbine generator performance is characterized by generator output and heat rate. Generator output is the electrical generation of the gas turbine generator measured at the generator terminals in SI (International System of Units) power units such as, watts, kilowatts and megawatts.

Heat rate is a measurement used in the energy industry to calculate how efficiently a generator uses heat energy. It is expressed in units of Btu (British Thermal Units) of heat required to produce a kilowatt-hour of energy (Btu/kWh). Heat rate is the ratio of the heat consumption to generator output. Heat consumption represents the thermal energy consumed by the gas turbine. Heat consumption is calculated as the product of the fuel mass flow rate and the heating value of the fuel. The convention, for gas turbines, is to use the lower heating value of the fuel. Consequently, heat rate for gas turbines is typically expressed on a lower heating value basis, and is calculated with the expression presented in Eq. 1 [5].

$$\text{Heat rate}\left(\frac{\text{Btu}}{\text{kWh}}\right) = \frac{\text{Heat consuption (Btu/h)}}{\text{Generator output (kW)}} \tag{1}$$

Operators of generating facilities can make reasonably accurate estimates of the amount of heat energy present in a given quantity of any type of fuel, so when this is compared to the actual energy produced by the generator, the resulting figure tells how efficiently the generator converts that fuel into electrical energy. Other main gas turbine performance parameters are listed in Table 1.

These parameters are commonly used in acceptance testing, testing to determine degradation of the machine, and operational range testing. Figure 4 shows the location where performance parameters of gas turbines can be measured [5].

The performance and reliability are governed by some standards such as those edited by The American Society of Mechanical Engineers (ASME), American Petroleum Institute (API) and International Organization for Standardization

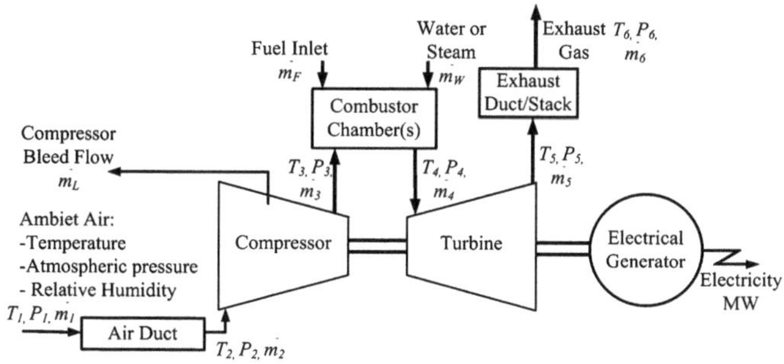

Fig. 4 Gas turbine generator performance parameters

(ISO). For performance analysis the ASME performance test codes (PTC) are used, and for reliability analysis the mechanical standards and codes are used.

3.1.1 Performance Test Codes

Performance testing is conducted on new gas turbines at the time of starting operation to determine compliance with its guarantee electrical output and heat rate. Performance testing is also conducted on existing units to determine their current electrical output and heat rate. ASME developed these performance test codes to provide explicit test procedures with accuracies consistent with current engineering knowledge and practice.

The objective of performance test codes is to provide standard directions and rules to conduct and to report results of tests of specific equipment and how to execute the measurement of related phenomena.

Over the past years, the international community has concentrated its efforts to spread the use of standards and codes produced by ASME, such as:

- ASME, Performance Test Code on Gas Turbines, ASME PTC 22 1997. The objective of the test(s) is to determine the power output and thermal efficiency of the gas turbine when operating at the test conditions, and correcting these test results to standard or specified operating and control conditions. Procedures for conducting the test, calculating the results, and making the corrections are defined, ASME PTC22, [1].
- ASME, Performance Test Code on Overall Plant Performance, ASME PTC 46 1996. The target of this code is establishing the overall plant performance, including combined-cycle and cogeneration power plants. The PTC 46 provides explicit procedures for the determination of power plant thermal performance and electrical output. Frequently the test results are used as defined by a contract for the basis of determination of fulfillment of contract guarantees.
- ASME, Performance Test Code on Test Uncertainly: Instruments and Apparatus PTC 19.1 1988. PTC 19 describes and specifies procedures for evaluation of

uncertainties in individual test measurements, arising from both random errors and systematic errors, and the propagation of random and systematic uncertainties into the uncertainty of a test result. With these parameters it is estimated an approximate test confidence level. This is especially important when computing guarantees in plant output and plant efficiency.

3.1.2 Mechanical Parameters

Regarding mechanical performance, the industry recognizes as the best standards those developed by the American Petroleum Institute (API) and The American Society of Mechanical Engineers (ASME). The ASME and the API mechanical equipment standards aid in specifying and selecting equipment for general petrochemical use. The main purpose of these mechanical standards is to provide specifications for the development of equipments with very high-quality, without losing the needs associated with the safety standards for the operation of gas turbines.

In the power generation industry it is becoming more frequent the use of heavy-duty gas turbines for open cycle and combined-cycle operation as a result of the short installation time and high efficiency coupled with low emissions. Additionally, the major gas turbine manufacturers develop increasingly flexible machines in relation to the type of fuel they consume, thereby increasing the available capacity of energy production. As a matter of fact, all of these features increase the installation and operation of plants to operate on the basis of market power.

The oil industry is one of the biggest users of gas turbine and uses that piece of equipment to generate power to move mechanical equipment such as pumps, compressors and electric generators. Therefore the specifications written by API and ASME are suitable for all areas, and operation and maintenance tips apply to all industries. Thus the specifications are well suited for the industry, and the tips of operation and maintenance apply for all industries, Boyce [7]. The main mechanical equipment standards are present in Table 2.

Table 2 only mentions the main mechanical standards. The application of the mechanical standards enables verification of the requirements for the operation of gas turbines. ASME and API developed these standards to somehow ensure high reliability of gas turbines, but among the rules it is not included a mechanism for calculating the reliability of that piece of equipment.

3.2 Gas Turbine Reliability and Availability

The terminology of reliability can sometimes be confusing, for example, most gas turbine manufactures (aero-derivative or heavy-duty) often claim that their products present highest reliability. This is why the term reliability must be clearly defined.

Table 2 Main mechanical equipment standards for gas turbines

Mechanical standard acronym	Mechanical standard title	Published year	Reaffirmed year
API Std 616	Gas turbine for the petroleum, chemical, and gas industry services	1998	
ASME B133.2	Basic gas turbine	1977	1997
ASME B133.7M	Gas turbine fuels	1985	1992
ASME B133.4	Gas turbine control and protection systems	1978	1997
ASME B133.8	Procurement standard for gas turbine auxiliary equipment	1981	1994
ASME B133.9	Measurement of exhaust emission from stationary gas turbine engines	1994	
ASME B133.5	Procurement standard for gas turbine electrical equipment	1978	1997
ASME B133.3	Procurement standard for gas turbine auxiliary equipment	1981	1994
API Std 618	Reciprocating compressors for petroleum, chemical, and gas industry services	1995	
API Std 619	Rotary-type positive displacement compressors for petroleum, chemical, and gas industry	1997	
API Std 613	Special purpose gear units for petroleum, chemical, and gas industry services	1995	
API Std 677	General-purpose gear units for petroleum, chemical, and gas industry services	1997	2000
API Std 614	Lubrication, shaft-sealing, and control-oil systems and auxiliaries for petroleum, chemical, and gas industry services	1999	
API Std 671	Special purpose couplings for petroleum, chemical, and gas industry services	1998	
ANSI/API Std 670	Vibration, axial-position, and bearing-temperature monitoring systems	1993	

3.2.1 Reliability Concepts

The formal definition of reliability is: "Reliability is the probability that a device will satisfactorily perform a specified function for a specified period of time under given operating conditions" [22]. (For more information see the Chap. 5). Reliability is represented by:

$$R(t) = 1 - F(t) \tag{2}$$

where:

$R(t)$ Reliability at time t

$F(t)$ Failure probability at time t

Based on this definition, reliability reflects the physical performance of products over time and is taken as a measure of their dependability and trustworthiness, Ohring [20]. Therefore, the highest reliability of a complex system, such a gas turbine, is really significant if calculated for an operational period (usually from 720 to 8700 h).

Moreover, some gas turbine manufacturers, define reliability as: "Probability of not being out of service when the unit is needed—includes forced outage hours (FOH) while in service, while on reserve shutdown and while attempting to start, normalized by period hours (PH)—percentage units (%)" [3].

Thus the reliability is calculated by:

$$\text{Reliability} = \left(1 - \frac{FOH}{PH}\right) \times 100 \tag{3}$$

where:

FOH Total forced outage (hours)
PH Operation period (hours)

Forced outage is related to the occurrence of random failures (unexpected). This type of failure is corrected with the corrective maintenance interventions.

The main difference between the ways of calculating reliability is: the manufactures use only downtime (*FOH*) and operation time (*PH*), without associating a probability distribution. When the operational time between failures is associated with a probability distribution, we obtain statistical measures of the failure probability what makes possible the use of more accurate tools for the analysis of equipment, extending the results in an operational period.

In the case of non-parametric reliability estimate (manufactures way), it is only known the punctual behaviour, and the result cannot be extended to the entire time domain. The major probability distributions associated with the reliability evaluation are exponential distribution, lognormal distribution and Weibull distribution. These distributions are those that best represent the behaviour of the reliability of electro-mechanical systems.

On the other hand, gas turbines are pieces of equipment composed of a large numbers of subsystems, with a large number of components, being considered complex systems. To analyze the reliability of complex systems the analyst can follow two main phylosophies, the quantitative and qualitative analysis. The most commonly used quantitative techniques are Markov chain, Monte Carlo simulation, block diagram analysis and fault tree analysis (for more information see the Chap. 5), and the qualitative techniques are Failure Modes and Effects Analysis (FMEA), Failure Modes, Effects and Criticality Analysis (FMECA), and Fault Tree Analysis (FTA).

To obtain satisfactory results, it is necessary to implement these techniques in a structured way, in other words, with the application of a method of analysis. The main methods of reliability analysis are presented in item 8.4.

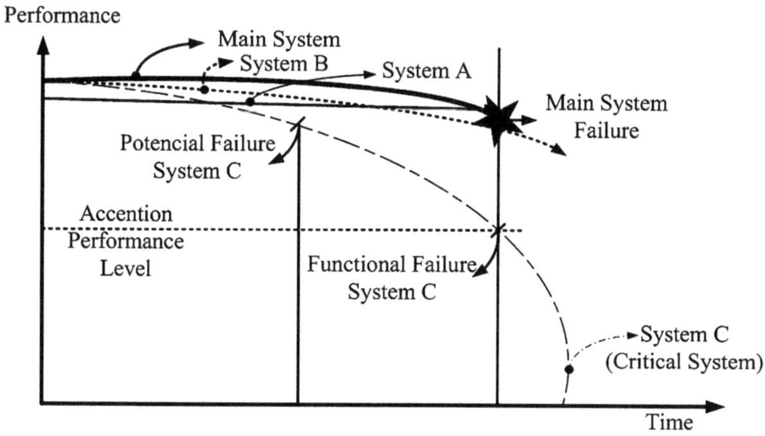

Fig. 5 Loss system performance caused by a functional failure of a component

A method for reliability analysis of complex system refers to the evaluation of the reliability of a system based on the reliability of its components. So it is necessary to define how the low reliability of components affects the system reliability. As an example, for gas turbine, when the bearings lubrication system, which plays a fundamental function, is presenting failure, the gas turbine has a forced outage. Figure 5 shows this behaviour for a system composed of 3 components (A, B and C). The performance of component C is reduced below the minimum necessary to fulfil its function (functional failure), leading to system failure, independently if the other components are presenting performance according to the operational standards.

The result of reliability analysis is usually the mean time to failure (MTTF). Suppose a number (n) of components that fail in service after successively longer times t_1, t_2, $t_3 \ldots t_n$. The MTTF is simply defined as:

$$\text{MTTF} = \frac{t_1 + t_2 + t_3 + \cdots + t_n}{n} \tag{4}$$

The importance of analyzing the reliability of each component is given by the close relationship between components reliability and system availability.

3.2.2 Maintainability

As well as satisfactory operational time is represented by the reliability, system behaviour during the repair time is represented by the maintainability. Maintainability (M) is formally defined as "a characteristic of design and installation which determines the probability that a failed equipment, machine, or system can be restored to its normal operable state within a given timeframe, using the prescribed practices and procedures" [6].

Therefore maintainability is the probability of maintaining a piece of equipment in a specified time duration. As a result of maintainability analysis it is possible to determine the mean time to repair (MTTR). The MTTR is simply defined as.

$$\mathrm{MTTR} = \frac{\text{Downtimes for repair } (t_r)}{\text{Number of repairs}} = \frac{t_{r1} + t_{r2} + t_{r3} + \cdots + t_{rn}}{n_r} \quad (5)$$

where t_{ri} is the time to repair after the ith failure and n_r is the number of repairs.

The downtime for repair (maintenance downtime), is a total time period during which the system is not in condition to perform its intended function. This time includes the mean active maintenance time, logistics delay time, and administrative delay time. In other words, maintenance downtime is strongly associated with maintenance planning strategies and maintenance policy [6].

In fact, the successful execution of a maintenance plan will reduce downtime and will increase equipment life avoiding the degradation of reliability and will be guaranteeing high availability.

3.2.3 Availability

Availability is defined as "the ability of equipment to successfully perform its required function at a stated instant of time or over stated period of time" [11] and is often calculated with the following equation:

$$A = \frac{\mathrm{MTTF}}{\mathrm{MTTR} + \mathrm{MTTF}} \quad (6)$$

The electric power industry uses the IEEE Std. 762 (Definitions for Use in Reporting Electric Generating Unit Reliability, Availability, and Productivity) to define the availability as: "availability measures are concerned with the fraction of time in which a unit is capable of providing service and accounts for outage frequency and duration" [14].

The IEEE Std. 762 was developed to aid the electric power industry in reporting and evaluating electric generating unit reliability, availability and productivity.

Among the most important indexes in the IEEE Std. 762, are the Availability Factor (AF) defined as "the fraction of a given operating period in which a generating unit is available without any outages", and Equivalent Availability Factor (EAF) defined as "the fraction of a given operating period in which a generating unit is available without any outages and equipment or seasonal deratings" [10].

The AF index, usually evaluated monthly, is reported in a Generating Availability Data System (GADS) and can be used for comparison between different generating systems. According to the standard the EAF is the suggested parameter to measure a plant Reliability, Availability and Maintainability (RAM) performance. The EAF factor is the equivalent percentage of time in a specified time period that the plant is capable of generating full-power output.

The EAF factor is less than 100% due to the occurrence of unplanned and planned outages.

The equivalent unplanned outage factor (EUOF) is the equivalent percentage of hours in a specified time period that the plant is not capable of operating at full-power because of unplanned outage events, and the equivalent planned outage factor (EMOF) is the equivalent percentage of hours in a specified time period that the plant is not capable of operating at full power because of planned outage events, usually associated with preventive maintenance planned intervention. That intervention includes tasks of inspection, testing or overhaul and is scheduled well in advance.

The EAF index is also dependent on how often a planned (or unplanned) outage occurs, on the length of time that an outage lasts once it has occurred, and on the loss of plant capacity associated with the outage.

For unplanned outages, the Std. IEEE 762 classifies the outages according to five classes, defined from class 0 to class 3 and maintenance. Unplanned outage Class 0 applies to a start-up failure and Class 1 applies to a condition requiring immediate outage. The unplanned outage Class 2 or Class 3 and Maintenance are determined by the amount of delay that can be exercised in the time of removal of the unit from service. That index is deterministic and can only be used for maintenance efficiency management.

In order to improve maintenance efficiency and to reduce maintenance costs it is recommended the use of reliability and maintainability concepts, calculated based on the power plant record of failure and operational context. The analysis allows the evaluation of MTTF and MTTR of the system, as proposed by the Eqs. 4 and 5.

As mentioned, the availability of a complex system, such as a gas turbine, is strongly associated with the parts reliability and the maintenance policy. That policy not only influences the parts repair time but also the parts reliability affecting the system degradation and availability.

Most of the maintenance tasks of power plant equipment are based on manufacturer's recommendations. Those recommendations are not always based on real experience data. Many manufacturers get very little feedback from the users of their equipment after the guarantee period is over. Fear of product liability claims may perhaps also influence the manufactures' recommendations.

In a large enterprise, such as a power plant, keeping asset reliability and availability, reducing costs related to asset maintenance, repair, and ultimate replacement are at the top of management concerns.

In response to these concerns a great number of maintenance planning methods have been developed including those based on the Reliability Centered Maintenance (RCM) concepts.

4 Reliability Analysis of Gas Turbine (Case Study)

A reliability analysis has as main objective to analyze the systems behaviour based on their "time to failure" database. These assessments are useful for planning activities that improve their operational times. The most used method is the Reliability Centered Maintenance. In the Chap. 6 the main concepts associated with that philosophy where analyzed.

In general the major objectives of RCM methodology are: Preserve functions; Identify failure modes that can defeat the functions; Prioritize function need; Select main monitoring systems to evaluate critical component degradation to allow the definition of maintenance actions before the occurrence of functional failure.

The sequence for implementation may vary depending on need and depth of analysis as discussed in Chap. 6.

4.1 Heavy-Duty Gas Turbine Case Study

This item presents a reliability analysis, using the RCM guidelines, of a heavy-duty F series gas turbine that has 150 MW ISO output (with natural gas as fuel).

The F series gas turbines are similar, despite the diversity of manufactures such as Siemens, Mitsubishi and GE. The F series are developed based on the concern with the problem of global warming which has been attributed to the burning of fossil fuels. At the same time its development was motivated to obtain high-efficiency with low-cost power generation.

The quality and the availability of fuels represent a continuous challenge to the manufacturers. The F series gas turbine, are designed to—through modifications in the combustor chambers—use a variety of fuels, increasing the system complexity from the point of view of reliability analysis.

A complex system reliability analysis consists on four main steps: functional tree construction; FMEA development; critical components selection; and maintenance policies selection, Carazas and Souza [9].

4.2 Functional Tree

The functional tree for heavy-duty gas turbine is presented in Fig. 6. The equipment is divided into five main subsystems: trunnion support, compressor, combustors, power turbine and start/stop subsystem.

Those main subsystems are divided into components, each one performing a specific function in connection with the subsystems main function. Loss of performance or functional failure in a component at the bottom of the tree can cause

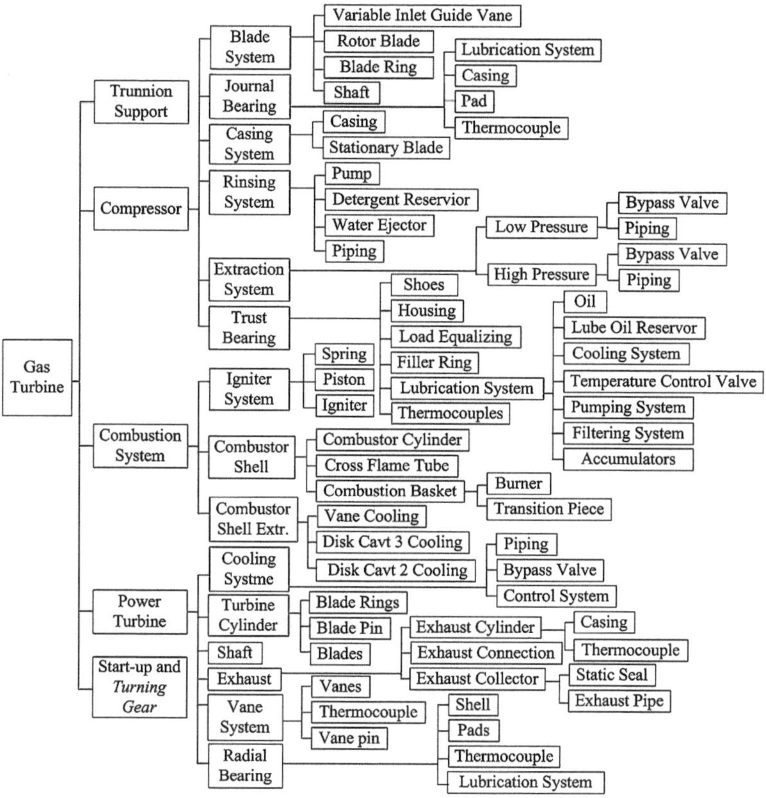

Fig. 6 Gas turbine functional tree

effects on all subsystems above it, causing a degradation in the gas turbine performance, represented by reduction in the nominal power output or degradation of some other operational parameter. If the gas turbine is operating in combined-cycle (as usually operate), the result of functional failure is more serious because it affects the power plant nominal output.

4.3 Failure Modes and Effects Analysis

In an objective manner, the FMEA is "Bottom up' approach where potential failure modes in every sub-item are analyzed, defining their effects on other sub-items, and their consequences on the system performance" [8].

Although there are many variants of FMEA, it is always based on a table, as shown in Table 3. In the left-hand column the component under analysis is listed; then in the next column the physical modes by which the component may fail are

Table 3 Failure Modes and Effect Analysis FMEA

Function	Failure mode	Failure causes	Failure effects	Critically
(A) Failure modes and effect analysis—Example: trunnion support				
Support turbine housing	Achieve ultimate limit state	Fatigue failure, fracture, Buckling	Loss of structural support, extensive damage to the turbine	9
	Achieve operational limit state	Plastic deformation due to overloading, existence of fatigue crack	Loss of structural stiffness, possible turbine vibration.	8
(B) Failure modes and effect analysis—Example: Electric motor				
Transform electrical energy in mechanical energy	Sheared shaft	Fatigue	Electric motor is locked	8
	Cracked housing	Fatigue/external shock or Vibration	Leakage of dust into motor/possible short circuit	6/8
	Bearing wear	Poor lubrication/ contamination/ overloading or high temperature	Noisy/heat build-up/ possible shaft locking	4/6/8

provided. This is followed, in the third column, by the possible causes of each of the failure modes. In the Table 3(A) and (B) are two examples of FMEA, for the trunnion support and electric motor respectively.

In the same way, the fourth column lists the effects of each failure mode that are classified according to the criticality scale.

The FMEA analyzes different failure modes and their effects on the system and classifies the level of importance based on "criticality" of the effect of failure. The ranking process of criticality can be accomplished by utilizing existing failure data or by a subjective ranking procedure conducted by a team of people with an understanding of the system operation. For the analysis of the level of severity of faults in power generation systems, the authors suggest using Table 4. This criticality scale expresses the degradation degree in the turbine operation.

The FMEA analysis was performed for each component listed in the end of a given branch of the functional tree.

The failure modes for the components were developed according to information from the literature [16, 7, 5, 8, 9]; from the catalogues of the manufacturers, [12, 13], Siemens 2006; [23], and based on the experience of the operators of various power plants, and study of maintenance records.

The analysis pointed out that the most critical components for the gas turbine, which are listed in Table 5.

Table 4 Criticality index description for FMEA analysis

Index	Effects on the turbine operation
1 (None)	This severity ranking is given when a component potential failure mode can cause reduction of performance of the equipment but does not cause damage to other eq. components, possibly affecting:- the eq. operation, when it is stopped to substitute or repair the failed component (substitution is not immediate);- no effect on power plant performance until the eq. is repaired
2 (Very minor)	This severity ranking is given when a component potential failure mode can cause reduction of performance of the eq. but does not cause damage to other eq. components, possibly affecting:- the eq. operation, when it is stopped to substitute or repair the failed component (substitution is not immediate);- very minor loss of power plant perf. until the eq. is repaired
3 (Minor)	This severity ranking is given when a component potential failure mode can cause reduction of perf. of the eq. but does not cause damage to other eq. components, possibly affecting:—the eq. operation, when it is stopped to substitute or repair the failed component (substitution is not immediate);—minor loss of power plant perf until the eq. is repaired;—possible minor effect in the environment
4 (Moderate)	This severity ranking is given when a component potential failure mode can cause reduction of perf. of the eq. but does not cause damage to other eq. components, possibly affecting:—the eq. operation, once it must be stopped to substitute or repair the failed component;—minor effect in the environment;—minor loss of power plant perf. until the eq. is repaired; minor loss of perf. in the power plant control system
5 (Significant)	This severity ranking is given when a component potential failure mode can cause reduction of performance of the eq. but does not cause damage to other eq. components, possibly affecting:—the eq. operation, once it must be stopped to substitute or repair the failed component;—the environment;—loss of power plant perf. until the eq. is repaired; loss of prf. in the power plant control system
6 (High)	This severity ranking is given when a component potential failure mode can cause reduction of prf. of the eq. but does not cause damage to other eq. components, possibly affecting:—the eq. operation, once it must be stopped to substitute or repair the failed component;—the environment;—the compliance with government requirements.;—loss of power plant prf. until the eq. is repaired; severe loss of prf. in the power plant control system
7(Severe)	This severity ranking is given when a component potential failure mode can cause unavailability of the eq. but does not cause damage to other eq. components, possibly affecting:—the eq. operation, once it must be stopped;—the environment in a severe manner;—the compliance with government requirements. The failure also causes the need of repair and/or replacement of the failed component. The plant is unavailable for short period of time

(continued)

Table 4 (continued)

Index	Effects on the turbine operation
8 (Very severe)	This severity ranking is given when a component potential failure mode can cause unavailability of the eq. but do not cause damage to other eq. components, possibly affecting:—the eq. operation, once it must be stopped;—the environment in a severe manner;—the compliance with government requirements. The failure also causes the need of repair and/or replacement of the failed component. The plant is unavailable for long period of time
9 (Hazardous effects)	This severity ranking is given when a component potential failure mode can cause severe damage to other components and/or to the eq., possibly affecting:—the eq. operation, once it must be stopped;—the environmental safety, including leakage of hazardous materials;—the safe power plant operation;—the compliance with government requirements. The failure also causes the need of repair and/or replacement of a great number of components. The plant is unavailable for long period of time

4.4 Reliability Analysis

Manufactures such as GE say they've obtained 100% reliability an 108-day period of continuous operation [13]. This value may not be used as a benchmark especially if the gas turbine is new. These reliability values are often used by manufactures, but also dependent on very stringent preventive inspections procedures and monitoring operation.

For example in Table 6 some data provided by manufactures related to inspection plans are presented and their effects on the calculated availability of a power plant are analyzed, [17].

Reliability can be defined as the probability that a system will perform properly for a specified period of time under a given set of operating conditions. Implied in this definition is a clear-cut criterion for failure, from which one may judge at what point the system is no longer functioning properly. For the gas turbine the failure criterion is any component failure that causes incapacity of generating the nominal power output.

Probably the single most used parameter to characterize reliability is the mean time to failure (or MTTF). It is just the expected or mean value of the failure time, expressed as:

$$\mathrm{MTTF} = \int_0^\infty R(t)dt \tag{7}$$

where:

$R(t)$ reliability at time t

T time period (h)

Table 5 Critical components of the gas turbine based on FMEA analysis results

System	Subsystem	Component	Failure mode	Criticality
Gas turbine	Structure		Reach ultimate limit state	9
			Reach operational limit state	8
	Compressor	Blades, Crown	Permanent deformation	8
			Rupture	8
			Presence of flaw	7
		Blades, fixture	Fatigue or overload rupture	7
			Permanent deformation	7
		Blades	Rupture	8
		Vanes	Loss of fixture	7
			Loss of geometrical tolerances	7
			Rupture	8
		Shaft	Rupture due to fatigue or overloading	8
			Permanent deformation	7
	Trust or guide bearing lubrication system	Oil tank	Rupture	9
		Oil pump	No outcome flow	7
		Electric Motor	Impossibility to convert mechanical into electrical energy	7
		Electrical control system	Electrical current interruption	7
			(Contato) Contactor: open circuit	7
			Contactor interrupts electric current incorrectly	7
			Interrupted electrical wiring	7
			Intermittent failure in electrical contact	7
		Piping	Rupture	8
			Transversal section clogging	8

(continued)

Table 5 (continued)

System	Subsystem	Component	Failure mode	Criticality
		Heat exchanger	Incapacity of cooling	7
			Partial heat exchanging capacity	7
		Filter	Filter clogging	7
	Air bleeding system	Low pressure valve	Incapacity of closure	7
			Incapacity of opening	7
		High pressure valve	Incapacity of closure	7
			Incapacity of opening	7
	Washing system	Valve	Incapacity of closure	8
			Leaking	7
			Partial closure	7
	Combustor system	Transition ring	Rupture	8
		Combustion chamber	Rupture	8
		Cross flame tube	Rupture	8
		Igniter system		8
		- Spring	Permanent deformation	8
		- Igniter	No supply power	8
			Rupture	8
	Cooling system	Flange	Rupture	8
		Piping	Rupture	7
	Tubine blade system	Blade	Rupture	8
		Blade ring	Permanent deformation	8
			Rupture	8
	Shaft	Couplings	Rupture	7
		Torque transmitter	Rupture	8
	Exhaust	Casing	Rupture	8
		Exhaust pipe	Rupture	8

Table 6 Manufacturer's inspections recommendations and expected availability

Year	Maintenance	Reliability %
First year	Periodic inspections	97
Year in which occurs	Combustion inspection	95
Year in which occurs	Hot gas path inspection	93
Year in which occurs	Major inspections	90

Random failures (represented by the exponential probability function) constitute the most widely used model for describing reliability phenomena. They are defined by the assumption that the rate of failure of a system is independent of its age and other characteristics of its operating history. In that case the use of mean time to failure to describe reliability can be acceptable once the exponential distribution parameter, the failure rate, is directly associated with MTTF, Lewis [15].

The constant failure rate approximation is often quite adequate even though a system or some of its components may exhibit moderate early-failures or aging effects, for example mechanical components. The magnitude of early failures is limited by strictly quality control in manufacturing and aging effects can be sharply limited by careful predictive or preventive maintenance. For example during transport and storage keep the parts in places with low temperature, low relative humidity, low solar irradiation and minimum vibration levels [4].

For a gas turbine in the beginning of its operational life, it is hard to affirm that the system present random failure, since its performance depends on commissioning and operational procedures and even on environmental conditions.

When the phenomena of early failures, aging effects, or both, are presented, the reliability of a device or system becomes a strong function of its age.

The Weibull probability distribution is one of the most widely used distributions in reliability calculations involving time related failures. Through the appropriate choice of parameters a variety of failure rate behaviours can be modelled, including constant failure rate, in addition to failure rates modelling both wear-in and wear-out phenomena.

In the sequence of the chapter, a reliability analysis of a gas turbine is presented. The database used in the reliability analysis corresponds to first 5 years of gas turbine operation. The reliability analysis is based on the time to failure data analysis. The turbine is modeled as one block. (See Chap. 5 or [15]).

Given that the gas turbine is composed mainly by mechanical parts, the reliability ($R(t)$) can be represented by a Weibull probability distribution, widely used in reliability calculations.

The two-parameter Weibull distribution, typically used to model wear-out or fatigue failures, is represented by the following equation:

$$R(t) = e^{-\left(\frac{t}{\eta}\right)^{\beta}} \tag{8}$$

Table 7 Weibull distribution parameters for gas turbine reliability calculations, Carazas and Souza [9]

System	Weibull distribution parameters
Heavy-duty gas turbine	$\beta = 0.58$ $\eta = 1014.24$

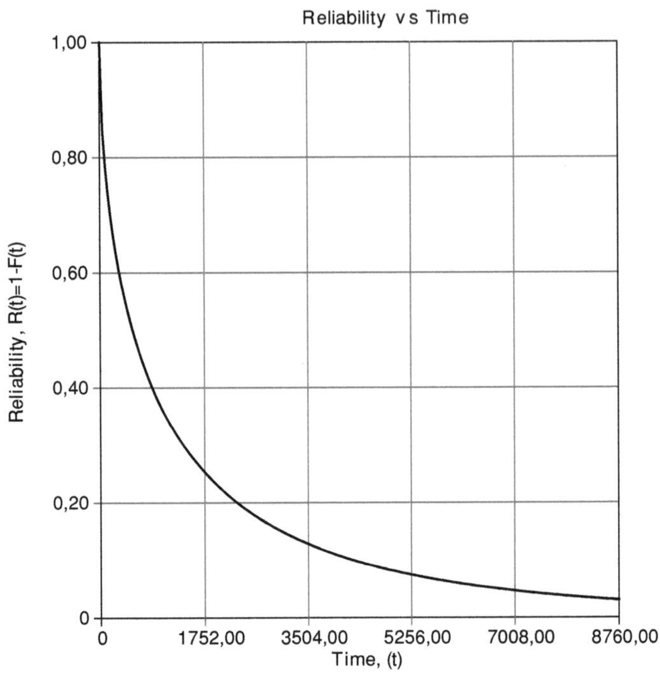

Fig. 7 Gas turbine reliability distribution

where:

$R(t)$	reliability at time t
T	time period (h)
β	Weibull distribution shape parameter Weibull distribution characteristic life (h)

The distribution parameters are estimated through the use of parametric estimation methods that fit the distribution to the 'time to failure' data. There are procedures for estimating the Weibull distribution parameters from data, using what is known as the maximum likelihood estimation method. For the gas turbine reliability analysis the software Weibull++ [21] was used for parameter estimation. Table 7 shows the Weibull distribution parameters for the heavy-duty gas turbines. The turbine reliability distribution is shown in Fig. 7.

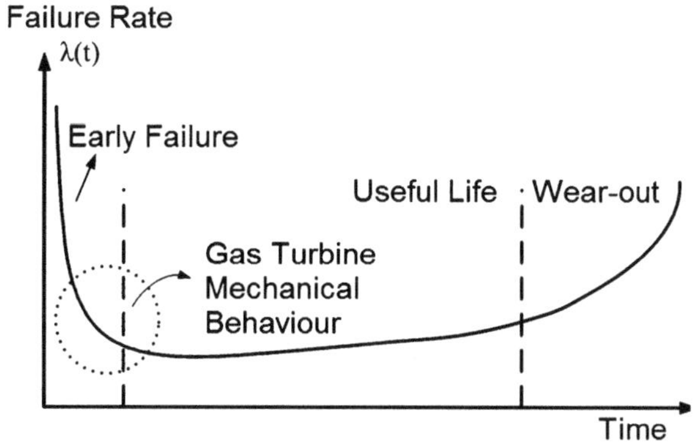

Fig. 8 Mechanical behavior of the turbine in the bathtub curve

The gas turbine has reliability distribution with shape parameters less than one. When $0 < \beta < 1$, the distribution has a decreasing failure rate. The gas turbine mechanical behaviour will follow the "failure rate" in the "bathtub curve". After the infant mortality phase, as shown in Fig. 8, it is expected that the equipment will present almost constant failure rate.

The reliability of gas turbine is equal 91.67%, calculated with Eq. 3, typically used by manufactures [3, 23].

The gas turbine under study presented 13 failures that caused equipment unavailability in the analysis period. Among those, more than five failures occurred in the first year operational. Most of them were related to high temperature in the combustors or excessive vibration on the bearings. The failure root-cause was sensor calibration problems. In the last 3 years there were two failures due to high temperature in the exhaust collector, caused by combustor failure.

For this reason most manufactures recommend doing boroscopy inspection after the first year of operation, or sooner, depending on the type of fuel. An effective boroscopy inspection can monitor the condition of internal components without the need for casing removal. Boroscopy inspections should be scheduled considering the operational context and environmental condition of the gas turbine and information from the O&M manuals.

The heavy-duty gas turbine designs incorporate provisions in both compressor casings and turbine shells for visual inspection of intermediate compressor rotor stages, turbine buckets and turbine nozzle partitions by means of the optical boroscopy. These provisions, consisting of radially aligned holes through the compressor casings, turbine shell and internal stationary turbine shrouds, are designed to allow the penetration of an optical boroscope into the compressor or turbine flow path area, [3]. Therefore the application of a boroscopy monitoring will assist the scheduling of power outages and preplanning of spare parts requirements, resulting in low maintenance costs and high availability and reliability of the gas turbine.

Table 8 Lognormal distribution parameters for gas turbine maintainability calculation, Carazas and Souza [9]

System	Parameters
Gas turbine	$\mu = 1{,}52$ $\sigma = 1{,}12$

Regarding the reliability analysis, the failures that may affect turbine availability were associated with components listed at the bottom of functional tree branches presented in Fig. 6 and were considered as critical in the FMEA analysis as shown in Table 5.

For the gas turbine, the early failure stage, defined by the great failure concentration in the first operational years, is mainly associated with the adjustment of control systems, mainly sensors.

4.5 Availability Analysis

To calculate availability it is necessary to estimate the parameters of the maintenance downtime (maintainability). The maintainability can be well described by a probability distribution. Typically the lognormal distribution is used to model the time to repair distribution of complex systems. The maintainability can be expressed as:

$$M(t) = \Phi\left(\frac{\ln t - \mu}{\sigma}\right) \tag{9}$$

where:

$M(t)$ maintainability at time t
μ lognormal distribution mean value
σ lognormal distribution standard deviation
$\phi(\bullet)$ standard normal distribution cumulative function

Based on the time to repair database for the gas turbine and using the software Weibull++ [21], the lognormal distribution parameters for maintainability modeling were calculated and are presented in Table 8.

The graphical representation of the maintainability probability distribution for gas turbine is presented in Fig. 9. The gas turbine has a mean time to repair equivalent to 8.56 h. The turbine has had simple failures, usually associated with sensors or control system devices that require a relative short time to repair.

For complex electrical–mechanical systems such as gas turbines, the determining factors in estimating repair time vary greatly.

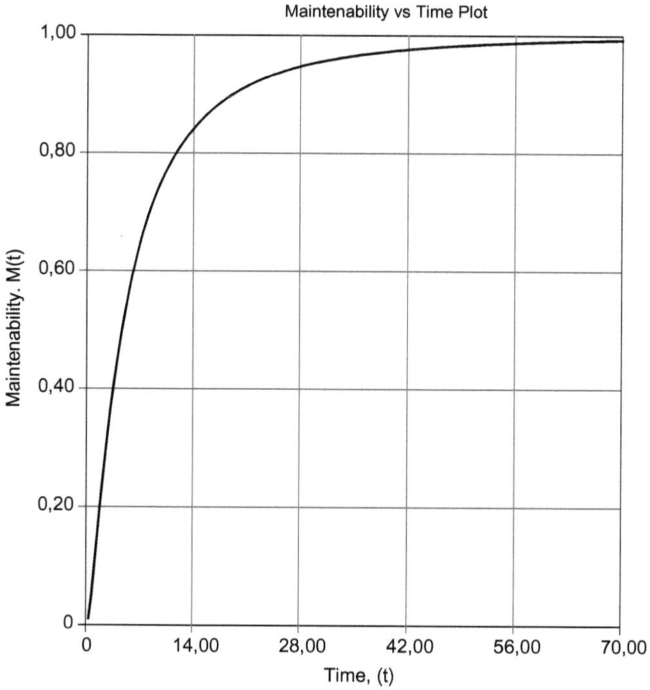

Fig. 9 Gas turbine maintainability distribution

In mechanical components, the causes of failure are likely to be quite obvious. The primary time entailed in the repair is then determined by how much time is required to extract the damage parts and install the new components. In contrast, if an electronic device (such as sensors or control units) fails, maintenance personnel may spend most of the repair procedure time in diagnosing the problem, for it may take considerable effort to understand the nature of the failure well enough to locate the part that is the cause. Conversely, it may be a rather straightforward procedure to replace the faulty component once it has been located.

Once the reliability and maintainability parameters are calculated the system availability can be estimated. Applying the Monte Carlo simulation method, the availability can be estimated for an operation time.

Considering the gas turbine operating over one year, corresponding to 8760 h, and using the reliability and maintainability probability distribution presented in Tables 7 and 8, respectively, the availability for the gas turbine is 99.35%.

The North American Electric Reliability Corporation [19] keeps available a reliability database based on North America power plants' performance that can be used as a benchmark for power plants availability analysis. According to that database the average availability of gas turbines with nominal output higher than 50 MW, within the period between 2002 and 2006, is 93.95%. The gas turbine

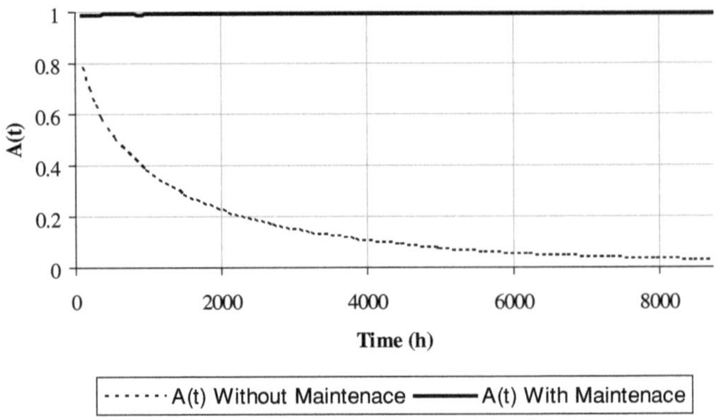

Fig. 10 Gas turbine availability simulation with maintenance

analyzed in the present study has higher reliability than the values presented in the NERC database. That comparison should be used only for initial evaluation of the gas turbine performance since that database does not clearly define the availability for heavy-duty gas turbines and the average age of the turbines used in the database are higher than the equipment evaluated in the presented paper. Nevertheless, the performance of the gas turbine analyzed in the present study can be considered satisfactory as far as the availability index is concerned.

Considering that the failure rate of the gas turbine is decreasing one can expect that the frequency of failures can continue to decrease until the equipment reaches the random failure stage. Based on the reduction of the failure frequency, caused by maintenance policies, routine inspections, or even operational procedure improvement, an availability increase can be expected. For example, with the execution of preventive inspection in periods of 3 months, the availability increases, as shown the simulation presented in Fig. 10. The first availability curve represents the availability of the turbine in normal operation, and the second represents the availability behavior with small peripheral equipment repairs and periodic inspections.

5 Influence of Fuel Flexibility in Gas Turbine Reliability (Case Study)

It is known that the constant changes in energy market (price fluctuations and the availability of fuels) can have a direct impact on the thermoelectric operational context. In order to face this reality, gas turbines that operate with a variety of fuels can be the key to success for power plant operators worldwide.

In this section it is presented a study conducted to analyze the reliability of a turbine with dual fuel system. The method of analysis varies if compared to

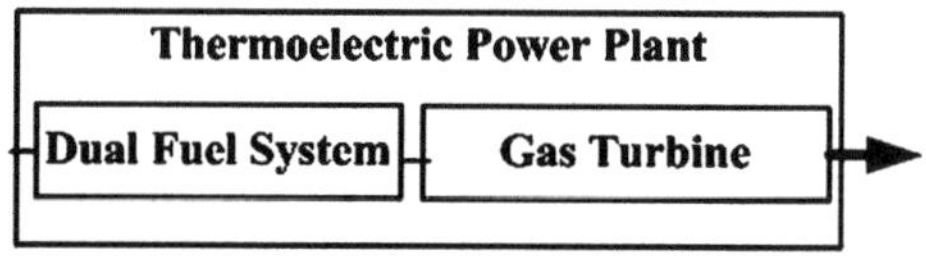

Fig. 11 Thermoelectric power plant reliability block diagram

item 8.4. The gas turbine is viewed as a block, with a particular reliability, and the dual fuel system as another block in series with the gas turbine block, as shown in Fig. 11.

5.1 Dual Fuel Systems

For any gas turbine-manufacturer, the fuels that will be used will have a great effect upon both the machine design and the materials of the components. Some gas turbine applications will always use highly refined and clean fuels; aircraft Jet engines are the prime example. In this case, materials and designs will primarily be limited by strength and oxidation characteristics. In most land-based gas turbines, however, use of cheaper, lower-grade fuels (such as heavy oil or Diesel oil) dictates that additional emphasis must be placed upon corrosion resistance, deposits, and the more challenging combustion characteristics.

For that reason, in case of adapting a turbine that uses natural gas to dual fuel, the combustor system must be replaced by a new design. These new combustors can burn a wide variety of fuels ranging from natural gas to the various processes gases, from naphtha to heavy residual oils. Dual fuel nozzles are often used to allow transfer between fuels without shutdown.

The major impact of the non-gaseous fuel properties on combustor design is on the liner metal temperature and carbon formation. The degree to which the fuel has been atomized is an important factor in establishing liner metal temperatures and reducing carbon formation. Usually in dual-fuel gas turbine the liquid fuel is air atomized at the fuel nozzle. Typical atomizing air pressure ratios (fuel nozzle air pressure/compressor discharge pressure) are in the range of 1.2–1.4 for light distillate fuels with higher ratios being required for heavy fuels. So, to transform a gas turbine in a dual fuel unit, a pressurized air system must be added to the auxiliary systems coupled to the turbine.

Special provisions are made for handling the ash-forming liquid fuels. These modifications are largely outside of the turbine core (which is largely standardized in construction) and are frequently external modular items. Examples are provisions for air atomizing, handling and treating the fuel, and cleaning the turbine to remove ash deposits.

Some ash-forming liquid fuels require a derating in firing temperatures, particularly in the larger machine models. This is accommodated by modifications in the control circuit, as opposed to changes in the gas turbine properly.

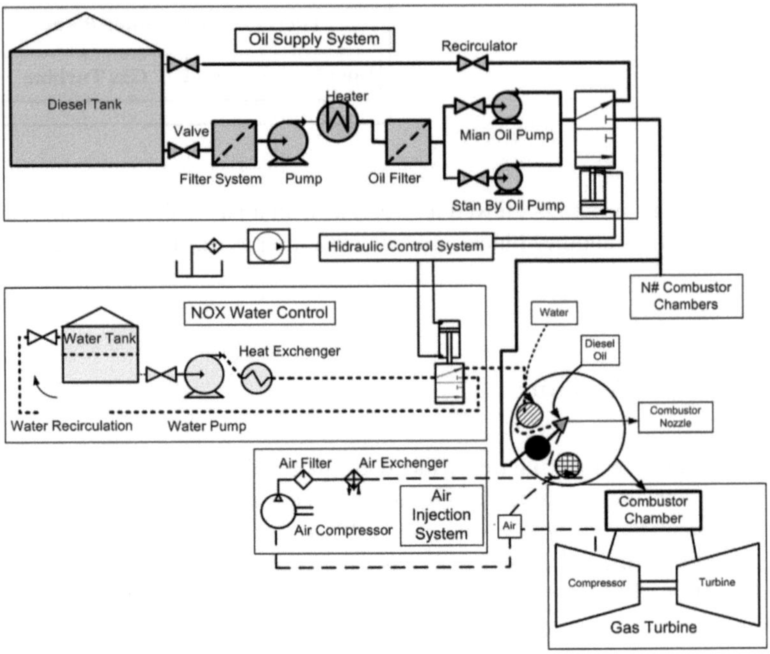

Fig. 12 Dual fuel flow diagram

Fuels may also require heating for pumping and forwarding. For heavy, lower-grade fuels, it may be necessary to heat the fuel to lower the viscosity to the operating range of the fuel transfer and filter systems. It may also be necessary to heat some crudes and heavy distillates to keep wax dissolved. Petroleum waxes occur to varying extents in crude oils depending on the geographical source, with the wax tending to become concentrated in the heavy distillate fractions.

Independent of the fuel type, there are certain precautions that can be taken at the plant site to minimize the chance of contaminants entering into the turbine. There are two approaches to ensure this condition:

- Minimizing the chance that the fuel will become contaminated by using careful transportation and storage methods;
- Removal of insoluble contaminants by setting filtration, centrifuging, electrostatic -precipitation, or a combination thereof.

To attain reliable combustor performance and to meet environmental restrictions for allowable NO_x, and other exhaust gas emission levels, it is necessary to use water in combustion, adding another auxiliary system to the turbine.

An illustrative flowchart of the auxiliary system added to a gas turbine in case of dual fuel conversion, considering the use of Diesel oil, is presented in Fig. 12.

The basic subsystems of those auxiliary systems are:

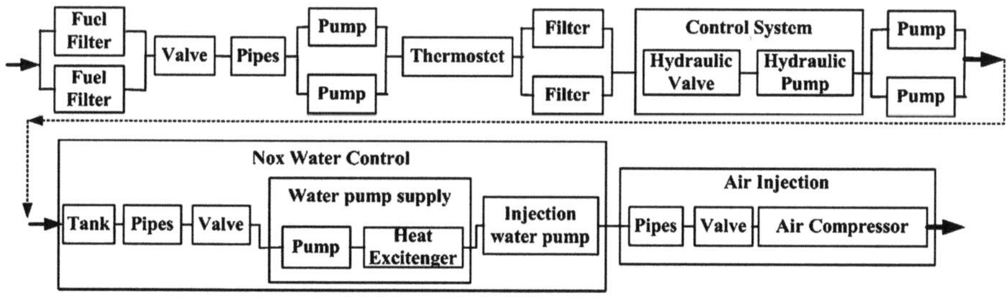

Fig. 13 Dual fuel reliability block diagram

- Oil supply system: fuel filters (including stand by filter), valves, piping system (oil line), fuel pump (including stand by pump);
- Oil heating system: piping system, valves, heater and control system (thermocouple);
- Staring system with oil: hydraulic control system (valves, hydraulic lines, pump), filter (including stand by system), oil valves, oil injection pump (including stand by pump);
- NO_x control system: water tank, water pump, water piping system, valves and heat exchanger;
- Atomizing air system: air filter, air piping system, compressor, valves, heat exchanger and atomizing air nozzle.

5.2 Reliability Analysis of Dual Fuel Systems

5.2.1 Reliability Block Diagram

The block diagram used to calculate the diesel oil system, which is added to the gas turbine in order to make dual fuel equipment, is presented in Fig. 13. That block diagram considers the possibility of gas turbine start up with diesel oil.

The reliability of each piece of equipment must be modeled to allow the system reliability estimative. The reliability of those pieces of equipment can be different in several power stations. The differences must be attributed to assembly conditions, operational differences and even variations in fuel quality.

5.2.2 Reliability Analysis

For the present analysis the authors will provide an estimate of the diesel oil supply system using the database published by Reliability Analysis Center (RAC) to evaluate the components reliability. Those data are presented in Table 9.

Table 9 Reliability of Dual Fuel system components

Component	Reliability distribution parameter
Oil supply system	
Fuel filter	Exponential distribution, $\lambda = 6.65 \times 10^{-5}$ failure/hour
Valves	Exponential distribution, $\lambda = 4.70 \times 10^{-8}$ failure/hour
Piping system	Exponential distribution, $\lambda = 4.7 \times 10^{-7}$ failure/hour
Fuel pump	Exponential distribution, $\lambda = 10.4 \times 10^{-6}$ failure/hour
Oil heating system	
Heater	Exponential distribution, $\lambda = 5.20 \times 10^{-6}$ failure/hour
Thermoset	Exponential distribution, $\lambda = 13.27 \times 10^{-6}$ failure/hour
Staring system with oil	
Hydraulic valves	Exponential distribution, $\lambda = 13.96 \times 10^{-6}$ failure/hour
Hydraulic pumps	Exponential distribution, $\lambda = 13.27 \times 10^{-6}$ failure/hour
Oil injection pump	Exponential distribution, $\lambda = 13.27 \times 10^{-6}$ failure/hour
Filter	Exponential distribution, $\lambda = 13.27 \times 10^{-6}$ failure/hour
NO_x control system	
Water tank	Exponential distribution, $\lambda = 2.24 \times 10^{-6}$ failure/hour
Piping system	Exponential distribution, $\lambda = 4.70 \times 10^{-7}$ failure/hour
Valves	Exponential distribution, $\lambda = 4.7 \times 10^{-8}$ failure/hour
Pump	Exponential distribution, $\lambda = 1.5 \times 10^{-5}$ failure/hour
Heat exchanger	Exponential distribution, $\lambda = 6.80 \times 10^{-6}$ failure/hour
Injection water pump	Exponential distribution, $\lambda = 1.5 \times 10^{-5}$ failure/hour
Atomizing air system	
Valves	Exponential distribution, $\lambda = 4.7 \times 10^{-8}$ failure/hour
Piping system	Exponential distribution, $\lambda = 4.7 \times 10^{-7}$ failure/hour
Compressor	Exponential distribution, $\lambda = 7.6 \times 10^{-7}$ failure/hour

All components reliability is modeled with an exponential distribution. That distribution is modeled to represent random failure modes. For mechanical equipment that approach is valid only if the component is submitted to a very stringent preventive or predictive maintenance policy in order to eliminate all aging dependent failure modes. The reliability distribution of diesel oil system is show in Fig. 14.

Considering the data presented in Table 9 and an operational period of 8760 h, the reliability of the Diesel oil system is 60.76%. The components that have the lowest reliability are the hydraulic components used in the startup system with Diesel oil. The Diesel oil pump in the fuel supply system also presents low reliability.

The hydraulic components reliability is significantly affected by the hydraulic oil quality. That quality is usually associated with the presence of particles in the fluid. Those particles can cause wear of valves and actuators affecting the hydraulic control system overall performance. The most traditional way to control hydraulic fluid contamination is by filtration and preventive change of the hydraulic fluid.

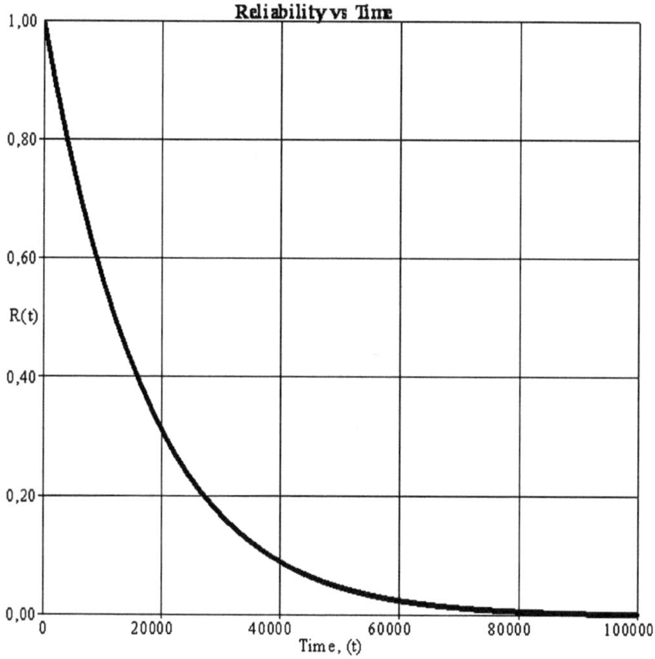

Fig. 14 Diesel Oil system reliability distribution

Taking in view that the Diesel oil subsystems can be considered as additional equipment that should be added to the standard gas turbine configuration, based on the plant operator choice, the gas turbine system reliability is influenced by those optional subsystem reliability.

6 Conclusion

This chapter presented the fundamentals of reliability analysis applied to gas turbines used in power plants. The traditional reliability and availability calculation method, based on index proposed by IEEE 762 [14], is compared with a method based on fundamental reliability concepts. The last one is strongly dependent on the existence of a maintenance and operation database where the operational times between failure and time to repair are correctly registered.

A heavy-duty gas turbine reliability analysis is presented and a discussion about the implementation of a dual fuel system is also presented. The reliability analysis seems suitable to identify critical components in the turbine and can be used as a guide to prepare predictive and preventive maintenance plans aiming at increasing turbine availability.

References

1. American Society of Mechanical Engineers (ASME) (1999) PTC 22 performance test code on gas turbine. ASME, New York
2. Siemens (2006) SGT6-5000F (W501F) 3 million hours fleet operational experience. In: Power-Gen International 2006, Orlando
3. Balevic D, Burger R, Forry D (2009) Heavy-duty gas turbine operating and maintenance considerations. GE Energy, Atlanta
4. Bajenescu T, Bâzu M (1999) Reliability of electronic components a practical guide to electronic systems manufacturing. Springer, Berlin
5. Black and Veatch (1996) Power plant engineering. Chapman & Hall, New York
6. Blanchard B, Verma D, Peterson E (1995) Maintainability a key to effective serviceability and maintenance management. Wiley, New York
7. Boyce M (2001) Gas turbine engineering handbook. Gulf Professional Publishing, Boston
8. Carazas F (2006) Availability analysis of gas turbines used in thermoelectric power plants. MSc Thesis presented at Polytechnic School—University of São Paulo (In Portuguese). São Paulo
9. Carazas F, Souza G (2009) Availability analysis of gas turbines used in power plants. Int J Thermodyn 12:28–37
10. Curley G (2006) Power plant performance indices in new market environment: IEEE standard 762 working group activities and GADS database. Power engineering society general meeting, IEEE, 1-4244-0493-2, p 5
11. Duffuaa S, Raouf A, Dixon J (1999) Planning and control of maintenance systems: modeling and analysis. Wiley, New York
12. GE Energy (2009) Heavy-duty gas turbine products. GEA-12985H (06/09), General Electric
13. GE Energy (2010) 6FA heavy duty gas turbine advanced technology for decentralized power applications. GEA17987 (07/2010), General Electric
14. Institution of Electrical and Electronics Engineers (IEEE) (2006) IEEE Std. 762 definitions for use in reporting electric generating unit reliability, availability, and productivity. IEEE, New York
15. Lewis E (1996) Introduction to reliability engineering. Wiley, New York
16. Lora E, Nascimento M (2004) Geração termeletrica planejamento, projeto e operação (in porgies). Interciência, Rio de Janeiro
17. Machado J, Gueiros Jr, Lima E (2008) Reliability analysis of ACH power plant in simple cycle and its impact on energy planning for the ONS (in Portuguese). In: Proceedings of the First petrobras themoelectric operational and Maintenance forum, Rio de Janeiro
18. Moubray J (1997) Reliability-centered maintenance. Industrial Press Inc, New York
19. North American Electric ReliabilityReliability Council (NERC) (2006) Generating availability report. available at www.nerc.com/~filez/gar.html. NERC
20. Ohring M (1998) Reliability and failure electronic materials and devices. Academic, San Diego
21. Reliasoft Corporation (2003) User's guide weibull++ version 6. Reliasoft Corporation, Tucson
22. Smith A, Hinchcliffe G (2004) RCM gateway to world class maintenance. Elsevier Butterworth-Heinemann, Oxford
23. Xia J, Antos R (2006) Siemens Power Generation SGT6-5000F(W501F) 3 Million hours fleet operational experience. Power-Gen International 2006, Orlando

Combined-Cycle Gas and Steam Turbine Power Plant Reliability Analysis

Gilberto Francisco Martha de Souza, Fernando Jesus Guevara Carazas, Leonan dos Santos Guimarães and Carmen Elena Patino Rodriguez

Abstract The reliability and availability of the combined-cycle thermal power plants depends on the perfect operation of all its systems (e.g. gas turbine, heat recovery steam generator, steam turbine and cooling system). The Heat Recovery Steam Generator (HRSG) is the link between the gas turbine and the steam turbine process having the function of converting the exhaust gas energy of the gas turbine into steam. In the cooling water system, heat removed from the steam turbine exhaust in the condenser is carried by the circulating water to the cooling tower, which rejects the heat to the atmosphere. This chapter presents reliability and availability analysis of a 500 MW combined-cycle thermal power station aiming at defining the most critical components of the main pieces of equipment as for power plant availability. The cooling tower cells are detailed evaluated in order to improve there availability through risk analysis and reliability centered maintenance concepts.

G. F. M. de Souza (✉) · F. J. G. Carazas
Department of Mechatronics and Mechanical Systems Engineering,
Polytechnic School, University of São Paulo, Av Prof. Mello Moraes 2231,
05508-900 São Paulo, Brazil
e-mail: gfmsouza@usp.br

L. d. S. Guimarães
ELETRONUCLEAR, Rua da Candelária 65, 20091-020 Rio de Janeiro, Brazil

C. E. P. Rodriguez
Department of Industrial Engineering, Engineering School,
University of Antioquia, Calle 67 # 53-108 Medellín, Colombia

G. F. M. de Souza (ed.), *Thermal Power Plant Performance Analysis*,
Springer Series in Reliability Engineering, DOI: 10.1007/978-1-4471-2309-5_9,

1 Introduction

The performance of any power plant can be determined by four elements:

(a) Capability to satisfy functional needs;
(b) Efficiency to effectively utilize the energy supplied;
(c) Reliability to start or continue to operate;
(d) Maintainability to quickly return to service after one failure.

It is evident that the first two measures are influenced by the design, manufacturing, construction, operation, and maintenance. Capability and efficiency reflect how well the power plant is designed and constructed. On the other hand, the last two measures, reliability and maintainability, are operations-related issues and are influenced by the power plant potential to remain operational and the efficiency of the repair aiming at returning to service process. It would be conceivable to have a power plant that is highly reliable, but does not achieve high performance.

Usually the power plant performance analysis is based on thermodynamic parameters evaluation in order to define the plant operational efficiency.

The ASME PTC 46 [1] code can be used to measure the performance of a plant in its normal operating condition, with all equipment in a clean and fully-functional condition. The analysis is based on key performance index and the most important one is the heat rate. This code provides explicit methods and procedures for combined-cycle power plants and for most gas, liquid, and solid fueled Rankine cycle plants. The scope of this code begins for a gas turbine based power generating unit when a heat-recovery steam generator is included within the test boundary. To test a particular power plant or cogeneration facility, the following conditions must be met. (a) means must be available to determine, through either direct or indirect measurements, all of the heat inputs entering the test boundary and all of the electrical power and secondary outputs leaving the test boundary; (b) means must be available to determine, through either direct or indirect measurements, all of the parameters to correct the results from the test to the base reference condition; and (c) the working fluid for vapor cycles must be steam. This restriction is imposed only to the extent that other fluids may require measurements or measurement methods different from those provided by this code for steam cycles. In addition, this Code does not provide specific references for the properties of working fluids other than steam. Tests addressing other power plant performance related issues are outside the scope of this code such as Reliability tests, which are tests conducted over an extended period of days or weeks to demonstrate the capability of the power plant to produce a specified minimum output level or availability. The measurement methods, calculations, and corrections to design conditions included herein may be of use in designing tests of this type; however, this code does not address this type of testing in terms of providing explicit testing procedures or acceptance criteria.

The most widely accepted definition of reliability is “the ability of an item, product, system, etc., to operate under designated operating conditions for a designated period of time or number of cycles”. The ability of an item to start or continue to operate can be designated through a probability (the probabilistic connotation), or can be designated deterministically.

The deterministic approach, in essence, deals with understanding how and why an item fails, and how it can be designed and tested to prevent such failure from occurrence or recurrence. This includes analysis such as review of field failure reports, understanding physics of failure, the role and degree of test and inspection, performing redesign, or performing reconfiguration. In practice, this is an important aspect of reliability analysis.

Reliability has then two connotations. One is probabilistic in nature; the other is deterministic. In this chapter, we deal with the probabilistic aspect of power plant operational performance.

It should be noted that reliability is also tightly related to operational safety, especially for nuclear power plants and other industrial applications that imply important risks for workers, public and environment. But these issues are not discussed within this chapter.[1]

The availability of a complex system such as a thermal power plant is strongly associated with the parts reliability and maintenance policy. That policy not only has influence on the parts repair time but also on the parts reliability affecting the system degradation and availability, [16].

Availability is a measure of percentage of time in which a plant is capable of producing its end product at some specified acceptable level. In a simple way, availability is controlled by two parameters, IEEE [12]:

- Mean time to failure (MTTF) which is a measure of how long, on average, the plant will perform as specified before an unplanned failure occurs, being associated with equipment reliability;
- Mean time to repair (MTTR) which is a measure of how long, on average, it will take to bring the equipment back to normal serviceability when it does fail.

Although reliability can be at least estimated during the plant design stages, its availability is strongly influenced by the uncertainties in the repair time. Those uncertainties are influenced by many factors such as the ability to diagnose the cause of failure or the availability of equipment and skilled personnel to carry out the repair procedures.

Eti, Ogaji and Probert [8] presented an approach for the integration of reliability concepts and risk analysis as guidance in maintenance policies for the Afam thermal power station. Those authors do not present the results expected or obtained with the application of those concepts.

A combined-cycle power plant is a combination of a fuel-fired turbine with a HRSG and a steam powered turbine. These plants are very large, typically rated in the hundreds of mega-watts. The use of combined-cycle power plants aims at using in the most efficient way the fossil fuels burned in turbine, typically natural gas or fuel oil, considered as nonrenewable energy resources.

---

[1] For reliability safety issues, see [11].

There are basically two general arrangements in combined cycle power plants, which are:

- Single-shaft: the combustion turbine and steam turbine drive a common generator in a tandem configuration, with only one HRSG. Single-shaft arrangement uses common systems for both turbines such as lubrication oil, simplifying the power plant auxiliary system complexity and maintenance planning.
- Multi-shaft: the combustion turbine and the steam turbine are coupled to a proper generator. Typically an arrangement with two gas turbines, two HRSGs and one steam turbine is used. Depending on the power requirements at the time, the multi-shaft combined cycle plant may operate only the fired turbine and divert the exhaust. However, this is a substantial loss of efficiency. Large fired turbines are in the low 30% efficiency range (although some manufacturers declares the possibility of operating heavy-duty gas turbines with efficiency close to 45%), while combined cycle plants can exceed 60% efficiency.

In the present chapter a multi-shaft combined-cycle power plant is analyzed.

The reliability and availability of the combined-cycle thermal power plants depends on the perfect operation of all its systems (e.g. gas turbine, heat recovery steam generator, steam turbine and cooling system).The HRSG is the link between the gas turbine and the steam turbine process, the function of HRSG is to convert the exhaust gas energy of the gas turbine into steam, Kehlhofer [13]. In the cooling water system, heat removed from the steam turbine exhaust is carried by the circulating water to the cooling tower, which rejects the heat to the atmosphere. Because of this direct path to the atmosphere, surrounding water bodies typically do not suffer adverse thermal effects. Cooling towers have been used for many years at power plants in locations where some water is available for cooling system use. The recirculating cooling water system arrangement incorporates an evaporative cooling tower as show in Fig. 1, [3].

2 Method of Reliability Analysis

The method first step consists in the elaboration of the functional tree for each piece of equipment installed in the power plant. The functional diagram allows the definition of the functional links between the equipment subsystems.

The next step is the development of the Failure Mode and Effects Analysis (FMEA) of each power plant equipment component in order to define the most critical components for plant operation. This criticality is based on the evaluation of the component failure effect on the system operation, [15]. For the definition of the system degradation, the FMEA analysis uses a numerical code, usually varying between 1 and 10. The higher the number the higher is the criticality of the component that must be evaluated for each component failure mode. For the plant analysis a criticality scale between 1 and 9 is proposed, [4]. Values between 1 and 3 express minor effects on the system operation and values between 4 and 6

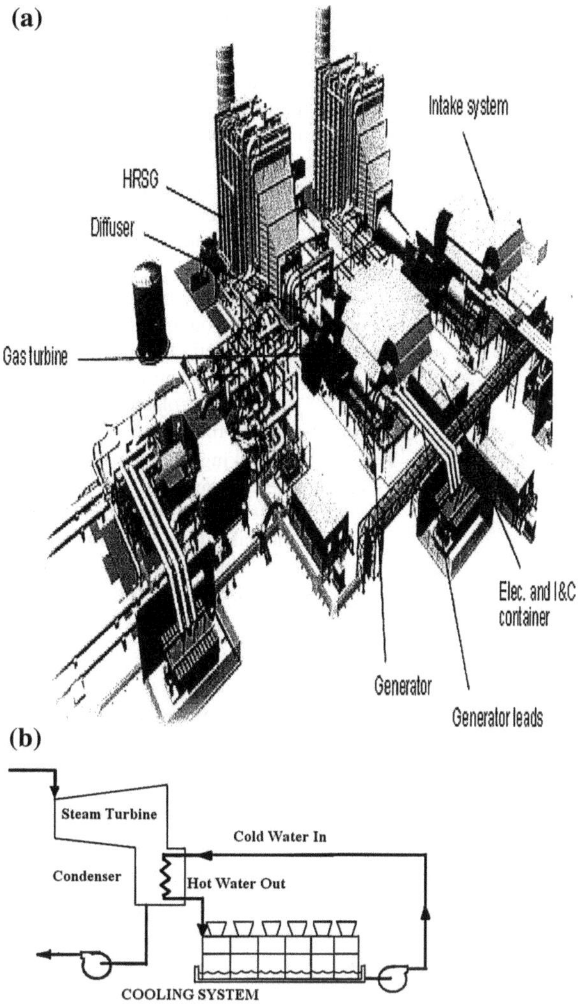

Fig. 1 Typical combined cycle thermal power plant general arrangement [3] **a** Power generation. **b** Cooling systemSystem

express significant effects on the system operation. Failures that cause the combined-cycle unavailability or environmental degradation are classified with criticality values between 7 and 9. These criticality values are shown in Table 1.

The method third step involves a reliability analysis. The failures should be classified according to the subsystem presented in the functional tree. The reliability of each subsystem is calculated based on the failure data base and the system reliability is simulated through the use of block diagram. Considering the 'time to repair' data and the preventive maintenance tasks associated with the equipment, the availability is evaluated using the block diagram.

Once the critical components are defined a maintenance policy can be proposed for those components, considering the RCM concepts. This maintenance policy philosophy has focus on the use of predictive or preventive maintenance tasks

Table 1 Criticality index description for FMEA analysis [4]

Criticality index	Effects on the turbine operation
7 (Severe)	This severity ranking is given when a component potential failure mode can cause unavailability of the equipment but does not cause damage to other equipment components, possibly affecting:–the equipment operation, once it must be stopped;–the environment in a severe manner;–the compliance with government requirements. The failure also causes the need of repair and/or replacement of the failed component. The plant is unavailable for short period of time.
8 (Very severe)	This severity ranking is given when a component potential failure mode can cause unavailability of the equipment but do not cause damage to other equipment components, possibly affecting:- the equipment operation, once it must be stopped;–the environment in a severe manner;–the compliance with government requirements. The failure also causes the need of repair and/or replacement of the failed component. The plant is unavailable for long period of time.
9 (Hazardous effects)	This severity ranking is given when a component potential failure mode can cause severe damage to other components and/or to the equipment, possibly affecting:- the equipment operation, once it must be stopped;- the environmental safety, including leakage of hazardous materials;–the safe power plant operation;–the compliance with government requirements. The failure also causes the need of repair and/or replacement of a great number of components. The plant is unavailable for long period of time.

that aim at the reduction of unexpected failures during the component normal operation, Smith and Hinchcliffe [16]. For complex systems, the occurrence of unexpected components failures highly increases maintenance costs associated with corrective tasks not only for the direct corrective costs (spare parts, labour hours) but also for the system unavailability costs.

So, the use of predictive or preventive tasks allows the programming of maintenance tasks in advance and also reduces the component failure probability during a given operation period and consequently increasing the system availability.

The reliability block diagram analysis not only allows the evaluation of the actual maintenance policy but also allows the prediction of possible availability improvement considering the application of new maintenance procedures, expressed by the reduction of corrective maintenance repair time.

In Fig. 2 a flowchart is used to explain the method's main steps [4].

3 Application

The method is applied on the analysis of a set of equipment installed in a 500 MW multi-shaft combined cycle thermoelectric power plant located in South America.

The plant uses two class F heavy duty gas turbines with nominal output close to 150 MW. The steam turbine, with three pressure levels is capable of generating nominal output higher than 200 MW.

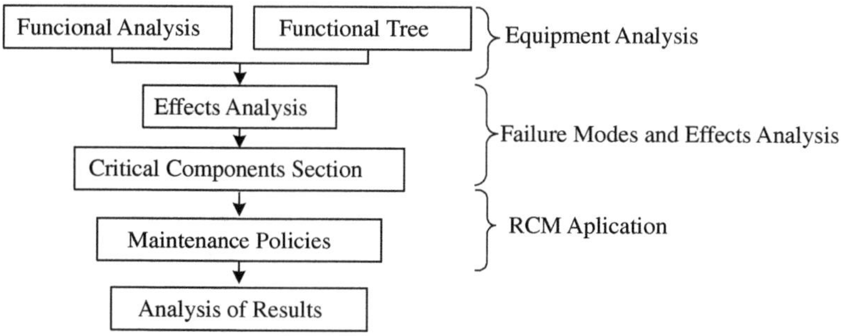

Fig. 2 Flowchart for complex system availability evaluation, Carazas and Souza [4]

The power plant HRSGs consist of three major components, which are Evaporator, Superheater, and Economizer. The equipment is classified as horizontal type HRSGs once the exhaust gas flows horizontally over vertical tubes, named harps. The equipment is also a triple pressure HRSG, presenting three sections: an LP (low pressure) section, a reheat/IP (intermediate pressure) section, and an HP (high pressure) section.

The power plant uses a condenser and a set of ten counter flow cooling towers that are independently operated.

Aiming at keeping the power plant output in the summer hottest days, an evaporative cooling system is installed in the inlet of both gas turbines aiming at reducing inlet air temperature and consequently controlling the turbine efficiency. The basic operation of the evaporative cooling system is based on the circulation of water through a heat exchanging media. Water is pumped from a tank to a header above the heat exchanging media. A spray system wets the top of the media. The water flows in the channels in the media, which are made of corrugated layers of fibrous material. The water flows down by gravity through the channels, wetting the material of the walls. The air absorbs the water which evaporates from the wall.

3.1 Functional Tree Development

The functional tree is developed for each piece of equipment installed in the power plant.

The functional tree for the gas turbine was presented in Chap. 8. In Figs. 3 and 4 the functional trees for the HRSG and steam turbine are respectively presented. In Fig. 5 the functional tree for the cooling system is presented, including condenser and cooling tower.

The main pieces of equipment of a combined cycle power plant, such as gas and steam turbine, HRSG and condenser have basic design characteristics but there might be some specific design features in the piece of equipment installed in a

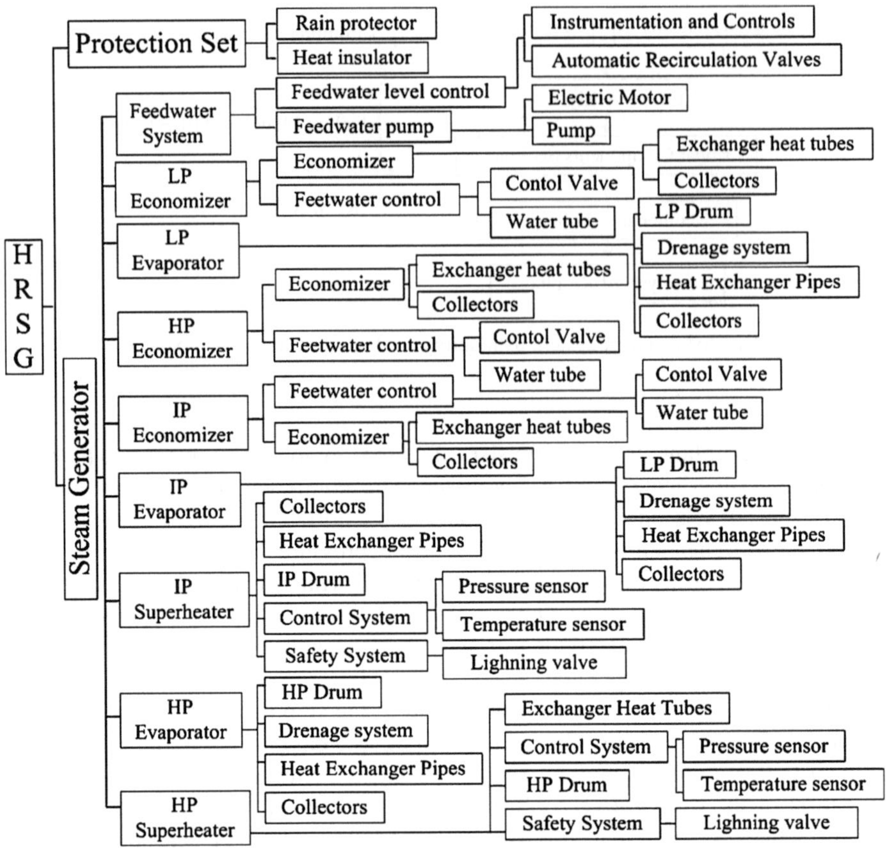

Fig. 3 HRSG functional tree, Carazas and Souza [6]

given power plant. So it is recommended that the functional tree must be developed for the equipment that composes the power plant under study.

Although all cooling tower possess essentially the same subsystems, such as circulating water pumps, circulating water piping and fans, there are differences between the technologies used by the manufacturers; therefore the functional tree must be developed for each specific cooling tower model. Specifically for the cooling tower under analysis, for each cell, the fan gearbox is located inside the cooling tower cell with the vertical output shaft below the fan.

The FMEA analysis was performed for each component listed in the end of a given branch of the functional tree. As for example, in Table 2 the critical components of the HRSGs are presented, Carazas, Salazar and Souza [6] and in Table 3 this analysis is executed for the cooling towers cells, Carazas and Souza [5].

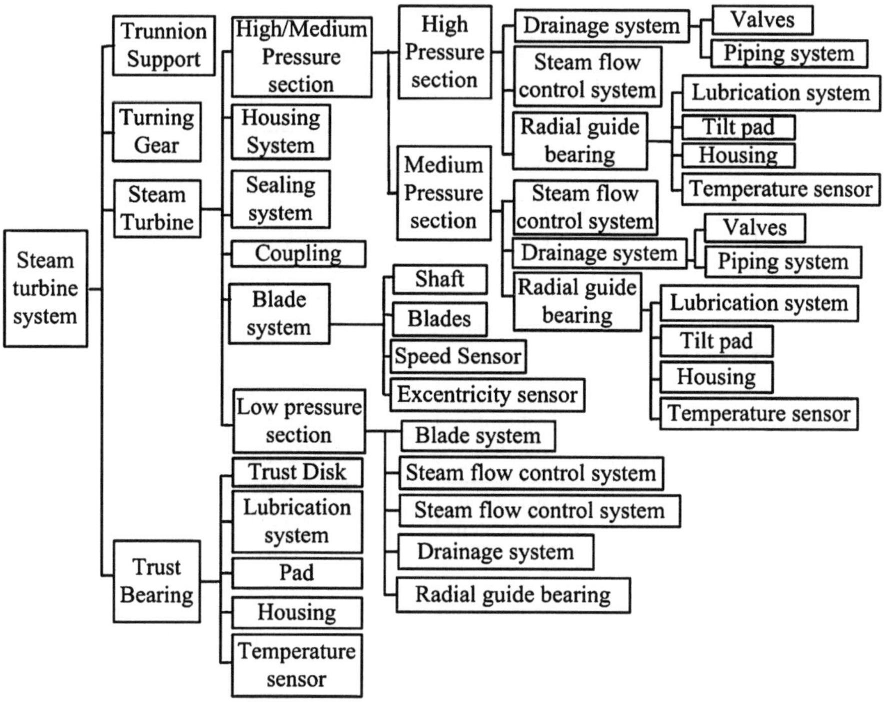

Fig. 4 Steam turbine functional tree

3.2 Reliability Analysis

The power plant must be modeled according to the reliability block diagram presented in Fig. 6. The power plant is a series system except for the cooling tower that is modeled as K out of N systems, meaning that it is necessary a given number of cooling towers units working (K) out of 10 to allow the plant to achieve nominal output.

Aiming at keeping the power plant nominal output during all year, the number of cooling towers units in operation must vary during the calendar year due to climate seasons.

The power plant site location presents low average temperature in winter (around 14°C), medium average temperature in autumn and spring (around 22°C) and high average temperature in summer (around 30°C). Based on the weather conditions, Table 4 presents the number of cooling tower units that must be operated to keep plant nominal output.

The cooling tower system operational strategy consists on using nine units and one unit is considered redundant equipment. During autumn and spring that operational strategy is being adopted by the plant operator. During winter, due to

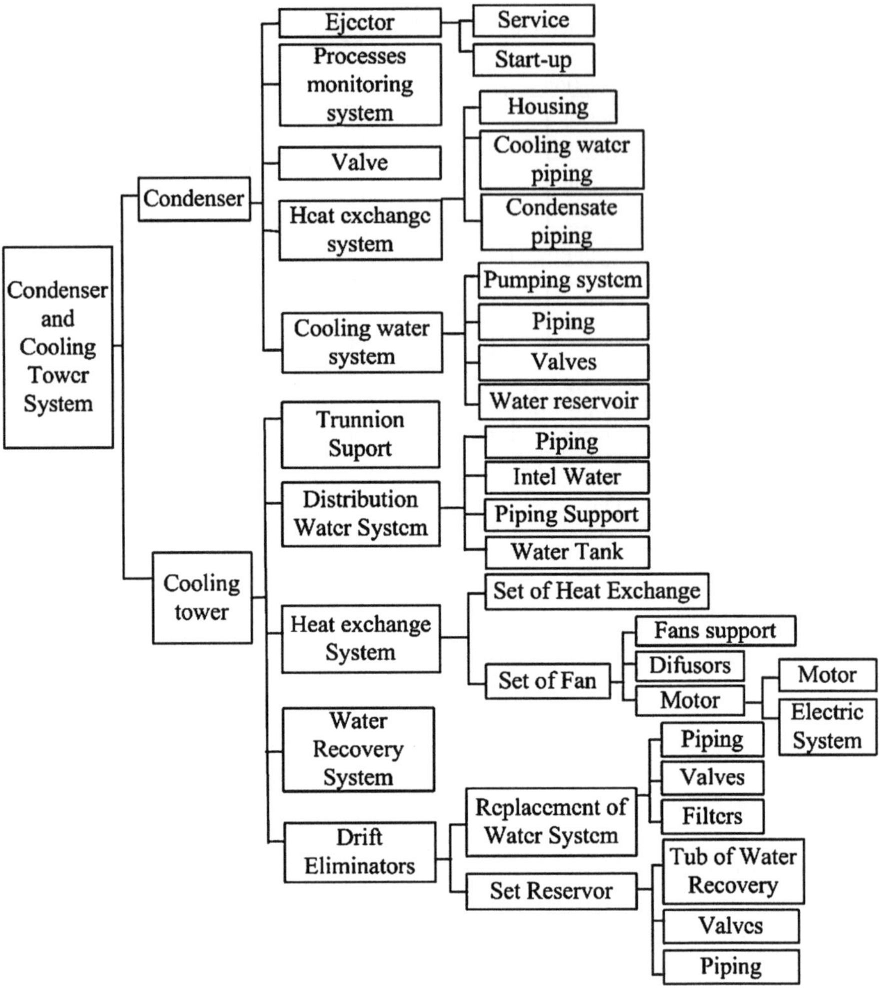

Fig. 5 Condenser and cooling tower functional tree, Carazas and Souza [5]

low air temperature, the cooling tower system can operate with 8 units without affecting the power plant nominal output.

During summer, the cooling tower system must operate with all ten units. The failure of any unit affects the power plant output. In some very hot days during summer it must be necessary to use an evaporative cooling system installed in the gas turbine inlet aiming at controlling turbine efficiency and heat rate to allow the power plant to deliver the nominal output.

The electrical generators coupled to each turbine are not considered in the present analysis which main focus is the evaluation of the influence of complex mechanical pieces of equipment in the power plant reliability. If those generators

Table 2 Critical components for HRSGs, Carazas, Salazar and Souza [6]

	Events	Critical components
1	Loss of the water flow in IP and HP evaporators	Feedwater pump. Water tube. Heat exchanger tubes (harps). Valves.
2	Dangerous increase in steam pressure inside Drums	Feedwater pump. Water tube. Heat exchanger tubes (harps). Control valves. Drainage system. Control system.
3	Increase in steam pressure in the HP system	Feedwater pump. Water tube. Heat exchanger pipes (harps). Control valves. Control system.
4	Loss of pressure control in the HP superheater	Water tube. Heat exchanger tubes (harps). Control system
5	Critical increase of the steam pressure	Water tube. Collectors. Heat exchanger tubes (harps). Control valves. Control system.
6	Rupture of the collectors and/or heat exchanger pipes	Collectors. Heat exchanger tubes (harps)
7	Feedwater pump failure	Check valve Electrical motor Pump

are considered in the analysis they should be represented by blocks added in series to the block diagram shown in Fig. 6.

The variation in the power plant configuration must be considered in the reliability analysis.

Reliability can be defined as the probability that a system will perform properly for a specified period of time under a given set of operating conditions. Implied in this definition is a clear-cut criterion for failure, from which one may judge at what point the system is no longer functioning properly. For the power plant analysis the failure criterion is any component failure that causes incapacity of generating the nominal power output.

The reliability analysis is performed for each of the pieces of equipment installed in the power plant, submitted to the same commissioning process and starting to operate at the same time. The reliability analysis is based on the time to failure data analysis.

Table 3 Critical components for cooling tower cell, Carazas and Souza [5]

Event	Component	Failure mode
Loss of cooling capacity	Structure support	Achieve ultimate limit state
	Distribution water system	
	Piping	Cross section blockage
	Inlet water	Cross section blockage
	Heat exchange system	
	Electric motor	No electric power
	Flexible shaft	Shaft cross section rupture
	Gear box	Gear tooth fatigue failure
		Shaft cross section rupture
	Coupling	Linkage between coupling and electric motor failure
		Coupling failure
	Water recovery system	
	Check valve	Incapacity to open
	Water piping	Cross section blockage

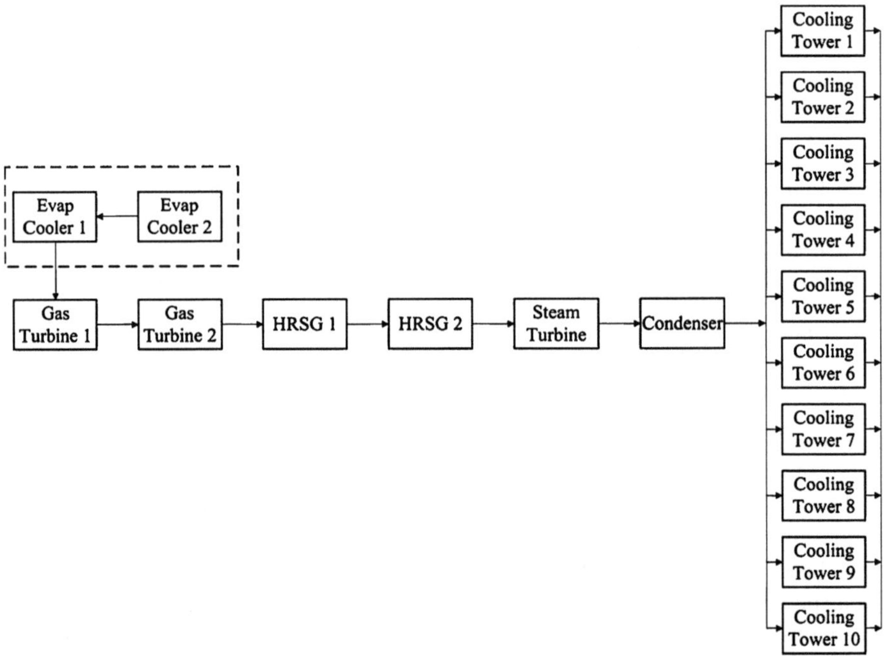

Fig. 6 Reliability block diagram for power station reliability analysis

For complex mechanical equipment such as gas and steam turbine, HRSG and condensers in the beginning of its operational life, it is hard to affirm that the system present random failure, since its performance depends on commissioning and operational procedures and even on environmental conditions.

Table 4 Cooling system configuration

Season	Number of cooling towers	Evaporative cooling system Operational
Summer	10	S[a]
Autumn	9	N
Winter	8	N
Spring	9	N

[a] For some specific weather conditions

When the phenomena of early failures, aging effects, or both, are presented, the reliability of a device or system becomes a strong function of its age. For that case the reliability is modeled through the use of Weibull probability distribution.

The gas turbines and HRSG reliability are modeled based on the use of the two-parameter Weibull distribution. The distribution parameters are estimated through the use of parametric estimation methods that fit the distribution to the 'time to failure' data. There are procedures for estimating the Weibull distribution parameters from data, using what is known as the maximum likelihood estimation method. The analysis of those systems is detailed presented by Carazas and Souza [4] and Carazas, Salazar and Souza [6].

The steam turbine and condenser reliability are also calculated based on ′time to failure′ database provided by the power plant. Taking in view that the technology associated with the design and manufacturing of those pieces of equipment are more mature than the design technology of heavy duty gas turbines, the reliability distribution is modeled with an exponential distribution.

For the cooling towers units and evaporating cooling systems the power plant failure database has not enough information as for reliability analysis. For those systems the reliability is calculated based on reliability database information, according to the studies present by Carazas and Souza [5].

3.2.1 Cooling Tower Unit Reliability

The cooling tower unit block diagram, for normal operation condition, is a series system using all subsystems present in the first level of the functional tree. Once the reliability of each component is defined, based on statistical analysis of their failure data, the cooling tower reliability is equal to the product of the subsystem reliability, as show in Fig. 7, and the system reliability is expressed by the Eq. 1.

$$R_{CT} = R_{SS}.R_{WS}.R_{HE}.R_{WR} \tag{1}$$

where R_{CT} is the reliability of cooling tower, R_{SS} is the reliability of support structure; R_{WS} is the reliability of distribution water system; R_{HE} is the reliability of Heat exchange system and R_{WR} is the reliability of water recovery system. Considering that each subsystem reliability can be modeled by an exponential

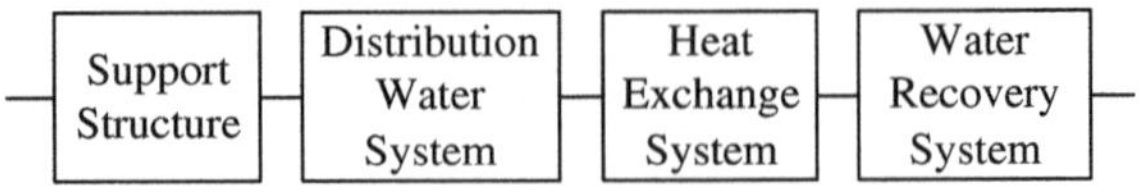

Fig. 7 Cooling tower reliability block diagram

distribution, the cooling tower reliability is also modeled by an exponential distribution which failure rate is calculated as:

$$\lambda_{CT} = \lambda_{SS} + \lambda_{WS} + \lambda_{HE} + \lambda_{WR} \tag{2}$$

where λ_{CT} is the cooling tower failure rate, λ_{SS} is the support structure failure rate; λ_{WS} is the distribution water system failure rate; λ_{HE} is the heat exchange system failure rate and λ_{WR} is the water recovery system failure.

The reliability of those subsystems can be estimated thought the following methods:

- Analysis of the historical failure database of the equipment.
- Analysis of the historical failure database of similar equipments.
- Analysis of prototypes reliability tests.
- Use of reliability prediction mathematical models based on commercial database.

For the present study, the selection of the most critical equipments as for reliability block diagram analysis is based on the failure database of the power plant. The critical components are those that present the greatest frequency of failure.

Unfortunately, the failure database does not clearly register the time between two consecutive failures in a given component that would support equipment reliability analysis. Thus, reliability estimate for the critical components is based on data book information.

The critical components are: support structure, electric motor, gear box and fans. The support structure is subjected to dynamic loading due to fan rotation. Electric motor and the gearbox are subjected to an environment with high humidity and subjected to dynamic loading due to fan rotation.

The simplified cooling tower unit block diagram is shown in Fig. 8.

Table 5 gives a list of the critical components that constitute cooling tower unit and the parameters of the reliability models, MIL-HDBK-217F [7] and Krishnasamy [14].

The cooling tower failure rate is 390×10^{-6} failures per hour.

3.2.2 Evaporating Cooling System

The evaporative cooler installation depends primarily on the plant operational characteristics and location. So the main goal of the present analysis is to evaluate the system reliability based on a standard configuration. That standard config-

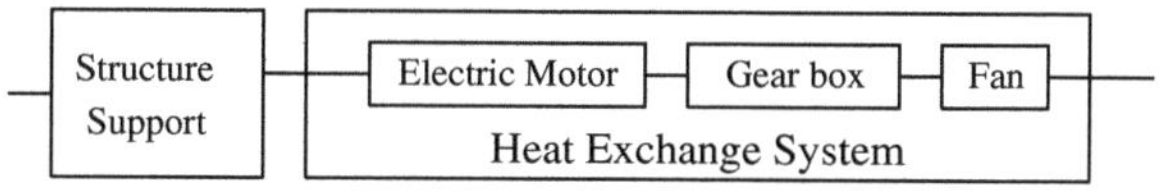

Fig. 8 Simplified cooling tower reliability block diagram

Table 5 Failure rate of cooling tower critical components

Subsystem	Component	Parameter
Structure support	Structure	$\lambda = 31{,}0 \times 10^{-6}$ failures/hour
Heat exchange System	Electric motor	$\lambda = 34{,}2 \times 10^{-5}$ failures/hour
	Gear box	$\lambda = 16{,}0 \times 10^{-6}$ failures/hour
	Fan	$\lambda = 1{,}20 \times 10^{-6}$ failures/hour

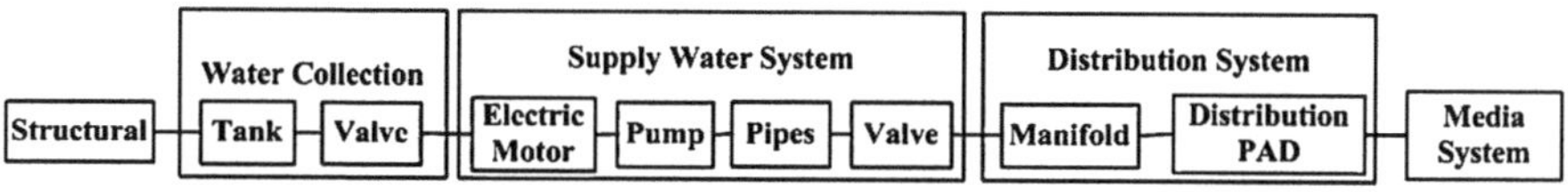

Fig. 9 Evaporative cooler system reliability block diagram

uration must represent the set of equipment that must be installed in any evaporative cooling system.

Taking in view the great number of pieces of equipment in the evaporative cooling system, the reliability cannot be calculated as only one part.

According to Sect. 2, for that kind of system the reliability block diagram is used to calculate system reliability. The proposed reliability block diagram for an evaporative cooling system is presented in Fig. 9. This block represents a generic evaporative cooler, considering the basic subsystems that any of those systems must present. The specificity of a given system must be represented by a new block added to the basic block diagram.

Analyzing the diagram presented in Fig. 9 it is possible to see that the evaporative cooling system is modeled as a series system, so the failure of one of the components will cause the failure of the subsystem. According to the reliability definition the subsystem failure does not only mean the total shut down of the cooling system but also the reduction of its performance. The use of reliability diagram only allows the evaluation of reliability but does not allow the evaluation of the system potential failure modes.

Nevertheless, the evaluation of reliability is important once it represents the chance that the evaporative cooling subsystem will be operational during a given operational period of the thermal power plant. This fact has a direct impact on the possibility that the gas turbine will achieve the predicted efficiency during a specific operational period.

To correctly estimate the evaporative cooling subsystem reliability for a given power plant it would be necessary to access the maintenance planning system

Table 6 Evaporative cooler subsystems reliability distribution

Component	Reliability distribution parameter
Structure	99% constant (estimated)
Water collection	
Tank	Exponential distribution, $\lambda = 2.24 \times 10^{-6}$ failures/hour
Valves	Exponential distribution, $\lambda = 4.7 \times 10^{-8}$ failures/hour
Water supply subsystem	
Pump plus electric motor	Exponential distribution, $\lambda = 10.40 \times 10^{-6}$ failures/hour
Piping system	Exponential distribution, $\lambda = 4.73 \times 10^{-7}$ failures/hour
Valves	Exponential distribution, $\lambda = 4.7 \times 10^{-8}$ failures/hour
Distribution system	
Manifold	Exponential distribution, $\lambda = 6.97 \times 10^{-6}$ failures/hour
Distribution pad	Exponential distribution, $\lambda = 4.7 \times 10^{-8}$ failures/hour
Media system	99%

database aiming at defining the time to failure probability distribution of each component listed in the block diagram. The reliability of those components will allow the evaluation of the evaporative cooling subsystem.

The goal of the present analysis is to estimate the reliability of a generic evaporative cooler and the components reliability can be initially estimated based on database information.

The most famous database as for thermal power plant equipment reliability analysis is the one provided by the North American Electric Reliability Corporation (NERC). That database is a collection of operating information associated with the performance of electric generating equipment, including reliability and availability data.

The database associated with gas turbine does not present any information regarding the failure of evaporative cooling subsystem. So, the reliability analysis for the subsystem presented on Fig. 9 will be based on data presented by the Reliability Analysis Center (RAC). Table 6 presents the probability functions and their parameters used to model the components reliability.

3.2.3 Power Plant Reliability and Availability Analysis

Table 7 shows the reliability and maintainability distribution parameters for each piece of equipment installed in the power plant.

In Table 7 β represents the shape factor of the Weibull distribution, η represents the scale factor of the Weibull distribution in hours, μ represents the mean of the Lognormal distribution, σ represents the standard deviation of the Lognormal distribution, both of them calculated in the logarithm scale, μ_n and σ_n represents respectively the mean and standard deviation of Normal distribution expressed in hours and λ represents the failure rate of the Exponential distribution expressed in failures/hour.

Table 7 Power plant pieces of equipment reliability and maintainability distributions

Equipment	Reliability distribution		Maintainability distribution	
	Distribution	Parameter	Distribution	Parameter
Gas turbine	Weibull	$\beta = 0.95$ η= 2,562.05	Lognormal	$\mu = 1.400$ $\sigma = 0.86$
Steam turbine	Exponential	λ= 0.0007	Lognormal	$\mu = 1.1395$ $\sigma = 2.022$
HRSG	Weibull	$\beta = 0.995$ η= 2,551.84	Lognormal	$\mu = 1.8932$ $\sigma = 0.9314$
Condenser	Exponential	λ= 0.0003	Lognormal	$\mu = 1.8684$ $\sigma = 1{,}5976$
Cooling tower (1 Unit)	Exponential	λ= 0.00039	Normal	$\mu_n = 8$ $\sigma_n = 1$
Evap Cooler	Exponential	λ= 0.000028	Normal	$\mu_n = 2$ $\sigma_n = 0{,}5$

For gas and steam turbines and HRSGs the reliability distribution reflects the preventive maintenance plan suggested by the manufacturer. For example, the gas turbine is subjected to special maintenance tasks according to the following schedule:

- After 8,000, 16,000, 32,000 and 40,000 operational hours the gas turbines are submitted to an inspection in the combustion system. The intervention takes around 7 days;
- After 24,000 operational hours the gas turbines are submitted to an inspection in the hot gas path. The intervention takes around 14 days;
- After 48,000 operational hours the gas turbines are submitted to a major maintenance. The intervention takes around 28 days.

In parallel to the gas turbine maintenance activities, the steam turbine and HRSGs are also submitted to complex preventive maintenance tasks with increasing complexity as a function of the operational hours. Both pieces of equipment are submitted to major maintenance after 48,000 operational hours.

In Fig. 10 the reliability of the power plant is presented for each climate season of the year. The difference between the reliability curves is caused by the increase in the number of cooling towers necessary to keep the nominal output of the power plant. The reliability curve named summer II is related to use of the evaporative cooling system installed in the air inlet of each gas turbine. It is possible to verify that the plant reliability decreases very fast along the operational hours.

The basic pieces of equipment of the power plant, such as the gas and steam turbines, HRSGs and condenser must always be operational to allow electric power generation. In Fig. 11 the reliability curve of the main equipment of the power plant (two gas turbines, two HRSGs and one steam turbine), considered as series system configuration, is presented. Through the comparison of Figs. 10

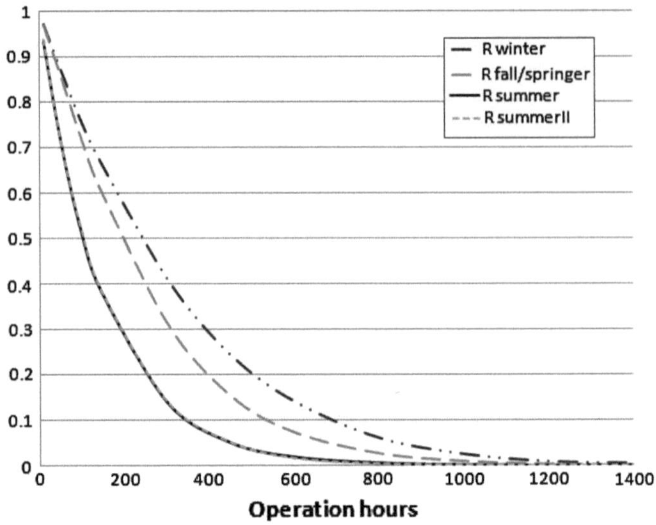

Fig. 10 Power plant reliability

Fig. 11 Gas turbines, steam turbine, hrsgs and condenser combined reliability

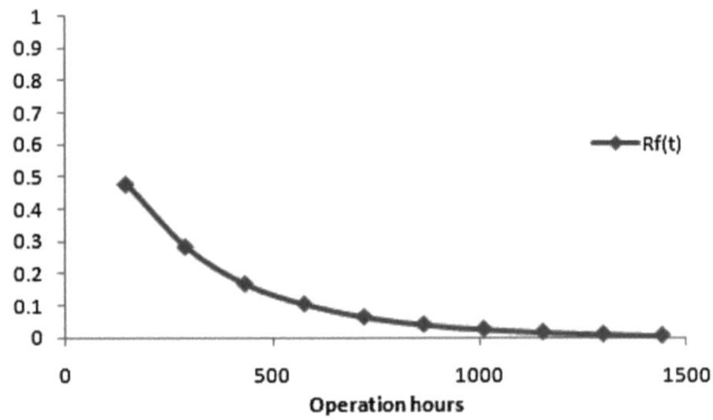

and 11 it is clear that the cooling tower reliability has great influence on the power plant reliability.

Although the main pieces of equipment have a very stringent maintenance policy, defined by the equipment manufacturers, some auxiliary systems that are not analyzed by the manufacturers can present failures that cause the performance reduction or even equipment shut down. As present by Carazas and Souza [4] and Carazas, Salazar and Souza [6], those auxiliary system must have their maintenance policy re-evaluated based on RCM concepts.

For the present analysis, the reliability of both gas turbines and HRSGs considers the improvement in the auxiliary system maintenance policy and the shape factor (β) of the Weibull distribution used to model their reliability is close to 1.

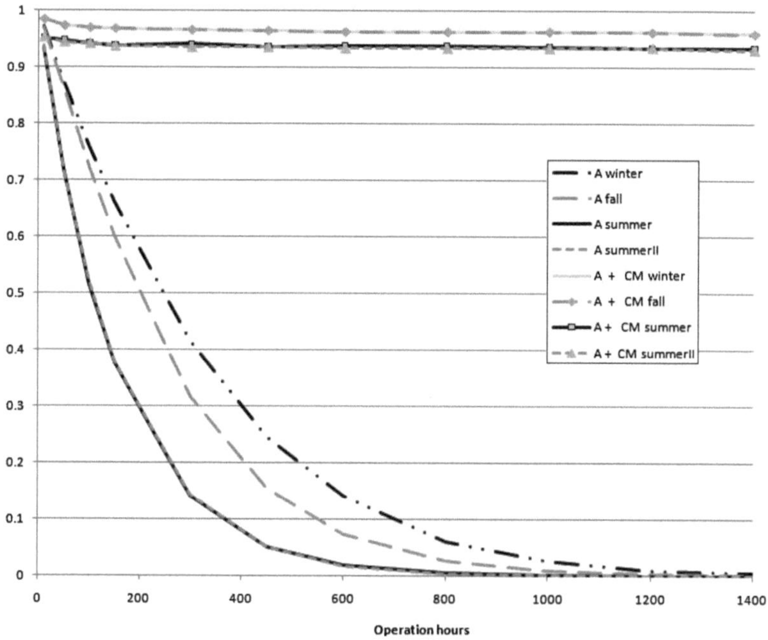

Fig. 12 Power plant availability

The steam turbine and condenser have their reliability modeled by exponential distributions indicating that their maintenance policy is considered sufficient to reduce the auxiliary systems failure frequency.

Another important performance index related to complex equipment efficiency evaluation is the availability index.

In Fig. 12 the availability index of the power plant is simulated considering the maintainability distributions presented in Table 7. The results are numerically presented in Table 8.

The power plant availability is calculated based on the relation between mean time to failure and mean time to repair, expressed as:

$$Availability = \frac{MTTF}{MTTF + MTTR} \tag{3}$$

The availability of the power plant is around 95% and that value is less sensitive to seasonal effects than the reliability index.

All calculations are made considering 1,440 h of continuous operational, corresponding to two months, once bimonthly the main pieces of equipment are subjected to predictive inspection aiming at detecting possible failure modes development.

The thermal power plant under analysis is normally used to complement the base generation of hydroelectricity. However there are seasons during the year,

Table 8 Comparison of power plant reliability and availability

Season	K	Reliability				Availability			
		Operational hours				Operational hours			
		168	336	720	1,440	168	336	720	1,440
Winter	8	0.5334	0.3233	0.1087	0.016	0.96	0.9587	0.9577	0.9554
Fall	9	0.5259	0.3054	0.0954	0.014	0.9563	0.9596	0.9568	0.9565
Spring	9	0.5259	0.3054	0.0954	0.014	0.9563	0.9596	0.9568	0.9565
Summer	10	0.273	0.0878	0.0072	0	0.9338	0.9302	0.9247	0.9226
Summer II	10 + EV	0.2794	0.0853	0.0067	0	0.9338	0.9344	0.9269	0.9235

usually in summer, that the hydro power plant lakes levels are low due to variations in the rainfall. By that time, the thermal power stations must generate the nominal output and any unexpected loss of performance can affect the power supply for some regions. During summer, the reliability of the power plant is strongly affected by the cooling towers units' reliability once all 10 units must be operational to provide plant nominal power output.

Furthermore, even with an incomplete failure database for the cooling towers, the plant unavailability data indicate a great number of failures in the cooling tower system.

Based on those results it is possible to select the cooling tower system as the most critical subsystem for the power plant. For that system a more complex analysis should be executed aiming at improving its reliability.

3.3 Cooling Tower System Reliability and Availability Improvement

The cooling tower unit can be considered mainly a mechanical system as for reliability and maintainability analysis. The operational conditions of the power plant, mainly associated with quality of recirculation water, and even the development of some specific failure modes of some components, such as fan unbalance, can affect the performance and degradation of some cooling unit components, mainly the gearbox. In reliability analysis this is named dependent failure modes.

The main forecasting methods of analysis previously outlined in the previous chapters of the book, allow highlighting the dependencies between failures. However, the relative difficulty of anticipating all these dependencies leads to multiply the approaches and to use more specific methods.

The advantages of these various methods, classified according to level of its effects on a set of interacting systems are summarized in Table 9, where a + sign gives a rough and qualitative indication of interest of each of these methods.

Table 9 The interest of predictive analysis methods in dependent failures, Guimarães [11]

	FMEA	FTA	BFCM	ETA
External initiators				
Internal initiators			+	+
Elementary systems				
Functional dependency			+	+
Common equipment dependency			+	+
Physical interactions			+	+
Human action dependency			+	+
Components				
Functional dependency	+	+	+	+
Common equipment dependency	+	+	+	+
Physical interactions	+	+	+	+
Human action dependency		+	+	+

The only method listed in the previous table that was not considered in the previous chapters is the Brief Failure Component Method (BFCM). The BFCM allows usually only for evidencing the simple failures and therefore must be completed by researching failure combinations that lead to undesired events. BFCM is the method that enables through inductive means, the determination of these failure combinations after carrying out FMEA. The set of abnormal operations (or failure modes) of a system is hence determined, Guimarães [11].

According to Guimarães [11], this method was created to analyse the safety in Concorde and Airbus aircraft. The research developed in nuclear fields contributed to the development of the method, notably theoretical fundamentals. The method is characterized by the introduction of certain specific concepts: brief internal failures, brief external failures, and brief global failures. The idea of "effects" allows for the connection between failures of abnormal operation, unwanted events, etc.

Numerous other specific methods have been developed for dependent failure analysis. These can be grouped into three main families:

(a) specific analyses of initiator events of dependent failures;
(b) analyses of generic causes; and
(c) analysis of operational experience

The specific analysis of initiator events of dependent failure aims to analyze in detail the effects of external initiators (aircraft collapse, earthquakes, etc.) or internal events (sizing accidents, loss of electrical panels, pipe ratchet ting, etc.) of the system, to limit the consequences and make them acceptable. It allows for the adequate sizing of elementary systems and the affected components.

The analysis of generic causes aim at predicting the occurrence and effects of common cause failures brought about by one or more generic causes. Among these analysis the following can be listed:

(a) prediction analysis of generic causes where the analyst seeks to identify system failures guided by the classification previously outlined; for the sources of potential failures, the analyst verifies the constructive provisions which enable the confrontation of these failures;
(b) zone analysis, that aim to analyze the dependencies between failures resulting from "geographical" location of certain components or subsystems; using:

- installation procedures
- physical array of components
- identification of possible failures; and
- identification of maintenance errors.

(c) analysis of human factors, which due to their importance are subject to special methods.

The detailed analysis of operational experience is a non-exhausting source of dependent failures. It requires a standardized and systematic gathering of all the incidents affecting systems and components. The wealth of this analysis depends very much on the quality of data gathering.

For the probability calculation of dependent failures and of common cause, two methods can be identified:

(a) explicit methods based on a precise knowledge of the causes of such failures, that allow the application of the general formula of conditional probabilities; and;
(b) parametric methods that are based upon statistical failure modelling, without research and listing of the causes.

Three parametric methods that deal, overall, with common cause failures can be indentified:

(a) parameter method β, Fleming, Kelley and Mosleh [9];
(b) multiple Greek letter method, Fleming, Mosleh and Deremer [10]; and
(c) shock method, Atwood [2].

The results of the operational experience analysis of nuclear power plant safety systems show the relatively high proportion of dependent failure incidents and the incidents that effectively occur. These demonstrate that the gains in availability or reliability achieved as a result of redundancies are inferior to those predicted theoretically. Therefore, for the more common components in safety systems, such as pumps, valves, engines, the gain can be estimated from 5 to 20 for a redundancy order of 2. An increase in redundancy by the order of 3 or 4 leads to very limited additional gains, maximum of 2–10. Beyond this, gains are marginal.

As an indication, and by no means exhausting the matter, a few generally applied preventive means for reducing the impact of these failures are, Guimarães [11]:

(a) During design phase:

- prevention of initiators that trigger failures of common cause, constructive dispositions for the control of:
  - environment and its aggressive factors
  - accidental environment originated within the installations, main plausible or hypothetical accidents considered in the design in a conservative manner.
- prevention of dependencies within elementary systems:
  - physical and geographical separation of redundant systems;
  - separation of safety functions assured by different systems;
  - functional diversity and system diversity;
  - different auxiliary systems;
  - allowance for periodical testing;
  - optimization of man–machine interface (automation of human actions that cannot be carried out with sufficient reliability within given time constraints; clear, precise and simulator proven, operational procedures, consideration of predicted human errors);
  - systematic research of dependencies through prediction analysis methods;
- prevention of the dependency between components:
  - physical and geographical separation of redundant systems;
- diversity of redundant components; (different design and manufacturing parties);
  - safety failure modes of the components
  - allowance for periodical testing;
  - systematic research of dependencies through prediction analysis methods;

(b) During Operational phase:

- prevention of dependent failures through systematic and detailed analysis of all the incidents and accidents upon installations and within similar installations
- prevention of human errors:
  - education, training and motivation of the operators, availability of several operators, incident diagnostics carried out by two independent and separate teams, and using separate means;
  - suspension of simultaneous maintenance carried out on important components, maintenance log carried out on important components controlled by other teams.

As for cooling tower analysis, once the system is already built, the plant operator can not modify the basic power plant design. Due to physical space problems it is not possible to increase the number of cooling towers units aiming at increasing the number of redundant units.

Another possibility is to change the cooling tower design aiming at reducing some components failures. Due to some inconsistencies in the power plant failure databases a more detailed failure analysis that would support any change in the unit is not presently feasible. In case of improvement in the database registers they should support more detailed FMEA analysis providing more information to define possible design changes.

The only possible way to reduce the frequency of failures in the cooling tower units is to improve the maintenance strategy.

3.3.1 RCM Concepts Applied to Cooling Tower Availability Improvement

Component manufacturers and suppliers tend to recommend a very conservative and costly maintenance approach. Changes in the power system market has led to a shift from technical to economic driving factors, including the maintenance planning with the aim of the increase of the operational lifetime and reduce costs.

Modern engineering systems are ideally designed for profitable operation throughout their service life in compliance with given requirements and acceptance criteria typically related to the safety of the personnel and the risk posed to the public and the environment. For ensuring this, it is necessary to control the development of deterioration processes by appropriate planning and performing of inspections and predictive maintenance actions. The predictive maintenance aims to reduce preventive maintenance tasks for critical components, this policy allows the reduction of unexpected failure occurrences that cause the system unavailability and are usually very expensive to repair.

The RCM philosophy supports the selection of maintenance tasks for critical components. In this way and based on the failure rates of these components, maintenance tasks are selected as shown in Table 10. The maintenance frequency is calculated to ensure a minimum reliability of 80% (for each critical component), calculated according to Eq. 2. The result of the analysis is displayed in the third column of Table 10.

A specific analysis can be developed for the cooling tower units' gearbox once this component usually presents a great number of failures due to the efforts imposed by the fan rotation.

The gearbox can present gear teeth failures due to transient overloads or even due to fatigue or even bearings failures due to problems with lubrication oil.

For this component it is recommended the use of condition monitoring based on vibration analysis. Based on velocity vibration measurements (usually expressed in mm/s RMS) the maintenance team can detect fan unbalance or misalignment, failure development in gear teeth and even early failure detection in rolling bearings. The vibration analysis could be complemented with oil analysis aiming at detecting the presence of debris that could be an indicative of gear tooth or bearing wear out. In Fig.12 it is presented the power plant availability considering the condition-based monitoring tasks applied to cooling tower units, without considering corrective tasks.

Table 10 Maintenance schedule for counterflow cooling tower, Carazas and Souza [5]

Description	Comments	Frequency of inspection activity
Overall visual inspection	Complete overall visual inspection to be sure all equipment is operating and safety systems are in place.	
Check tower structure	Check for loose fill, connections, leaks, etc./inspect and readjust the unions in case they have lost the adjustment due to vibration. Inspect the presence of cracks or deformations in the structure.	Monthly/
Bimonthly		
Fan electric motor condition	Check the condition of the fan motor through temperature or vibration analysis and compare to baseline values.	Monthly
Check fan blades	Check for excessive wear and secure fastening	Monthly
Flexible shaft	Check the condition of the flexible shaft fan through temperature or vibration analysis and compare to baseline values.	Monthly
Check motor supports	Check for excessive wear and secure fastening	Monthly
Motor alignment	Aligning the motor coupling allows for efficient torque transfer	Monthly
Check drift eliminators, louvers, and fill	Look for proper positioning and scale build up	Monthly
Inspect nozzles for clogging	Make sure water is flowing through nozzles in the hot well	Annually
Check bearings	Inspect bearings and drive belts for wear. Adjust, repair, or replace as necessary.	Annually
Motor condition	Check the condition of the motor through temperature or vibration analysis to assure long life.	Monthly
General recommendations for predictive and preventive maintenance		
Vibration	Check for excessive vibration in motors, fans, and pumps	
Test water samples	Test for proper concentrations of dissolved solids, and chemistry. Adjust blowdown and chemicals as necessary.	
Check lubrication	Assure that all bearings are lubricated per the manufacturers' recommendation.	
Clean tower	Remove all dust, scale, and algae from tower basin, fill, and spray nozzles.	
Piping	Check the leaks or excessive corrosion. Monitor the pressure of operation of the system to avoid very high pressures, and inspect the filter system to prevent the entry of corrosive agents	
Thermographic Analysis	Check and monitoring motors, bearing and pumps	

4 Conclusions

This chapter presented a reliability-based analysis of a thermal power plant aiming at detecting the critical components as for its long-term reliable performance.

According to the standards associated with in field performance analysis of thermal power plants, the reliability evaluation tests can be performed during plant commissioning but the codes do not specify how to develop those tests and how to analyze and to present the tests results.

The reliability and availability of a combined-cycle thermal power plant is presented focusing on defining the reliability distributions of plant main pieces of equipment. Those distributions were defined based on failure data recorded in the power plant maintenance record.

The cooling tower units are selected as the most critical pieces of equipment considering the power plant operational profile. For that piece of equipment some changes in the maintenance plan are suggested based on the RCM philosophy.

In order to increase the use of reliability concepts in thermal power plant performance analysis it is expected that the presence of regulatory framework can have a determining influence on the use of risk and reliability methods in practice.

References

1. American Society of Mechanical Engineers (ASME) (1996) PTC 46 performance test code on overall plant performance. ASME, New York
2. Atwood CL (1984) User's guide to binomial failure rate, NUREG-CR-2729. U.S. Nuclear Regulatory Commission, Washington
3. Black and Veach (1996) PowerPower plant engineering. Chapman and Hall, New York
4. Carazas F, Souza G (2009) Availability analysis of gas turbines used in power plants. Int J Thermodyn 12:28–37
5. Carazas FJ, Souza GFM (2009) Method for cooling towers maintenance policy selection based on rcm concepts. Proceeding of the 20th International Congress of Mechanical Engineering (COBEM 2009), Gramado, p 10
6. Carazas FJG, Salazar CH, Souza GFM (2010) Availability analysis of heat recovery steam generator used in thermal power plants. Energy. doi: 10.1016/j.energy.2010.10.003
7. Department of Defense (DoD) (1991) MIL-HDBK-217F Reliability prediction of electronic equipment. Department of Defense, Washington
8. Eti M, Ogaji S, Probert S (2007) Integrating reliability, availability, maintainability and supportability with risk analysis for improved operation of the AFAM thermal power-station. Appl Energy 84:202–221
9. Fleming KN, Kelley AP Jr, Mosleh A (1983) On the analysis of dependent failures in risk assessment and reliability evaluation. Nucl Saf 24:637–657
10. Fleming KN, Mosleh A, Deremer RK (1986) A systematic procedure for the incorporation of common cause events into risk and reliability models. Nucl Eng Des 93:245–273
11. Guimarães LS (2001) Risk management and system safety. iEditora, São Paulo
12. Institute of Electrical and Electronics Engineers (IEEE) (2006) Std. 762 standards definitions for use in reporting electric generating unit reliability, availability, and productivity. IEEE, New York

13. Kehlhofer RH, Warner J, Nielsen H, Bachmann R (1999) Combined-cycle gas and steam turbine power plants. PennWell Publishing Company, Tulsa
14. Krishnasamy L, Khan F, Haddara M (2005) Development of risk-based maintenance rbm strategy for a power plant. J Loss Prev Process Indust 18:69–81
15. Lewis E (1987) Introduction to reliability engineering. Wiley, New York
16. Smith A, Hinchcliffe G (2004) Reliability-centered maintenance: a gateway to world class maintenance. Elsevier Butterworth-Heinemann, New York

Risk-Based Inspection and Maintenance (RBIM) of Power Plants

Faisal Khan, Mahmoud Haddara and Mohamed Khalifa

Abstract The present chapter presents the basic concepts associated with Risk-based Inspection and Maintenance (RBIM) philosophy and their application in maintenance planning aiming at controlling power plant equipment degradation. The basic steps of the method are described, such as inspection sampling, inspection planning and maintenance activity selection based on degradation mechanism evolution, risk assessment and optimization of maintenance plan. The method is customized for power plant analysis considering the constraints associated with that application. Two case studies are presented: the first one is related to a pipeline analysis and the second one is a complete analysis of a power-generating unit.

1 Introduction

1.1 Deterioration Mechanisms

Processing assets and structures used in power plants are subjected to several structural deterioration mechanisms during their operation. Structural deterioration may result in damage, deformation, defects or performance degradation. The most common deterioration mechanisms are:

F. Khan (✉) · M. Haddara · M. Khalifa
Process Engineering, Faculty of Engineering and Applied Science,
Memorial University, St John's, NL A1B 3x5, Canada
e-mail: fkhan@engr.mun.ca

G. F. M. de Souza (ed.), *Thermal Power Plant Performance Analysis*,
Springer Series in Reliability Engineering, DOI: 10.1007/978-1-4471-2309-5_10,

1.1.1 Corrosion

Results in metal loss, pitting, cracks or/and degradation of material properties due to changes in the material microstructure. Corrosion could be general or localized. General corrosion is a metal loss widely distributed over the surface area of an asset. Localized corrosion results in a localized metal loss or cracks at different small areas over the material surface such as pitting corrosion, galvanic corrosion, crevice corrosion, stress corrosion cracking and Hydrogen induced cracking.

1.1.2 Fatigue

Fatigue occurs due to fluctuating stress (mechanical fatigue) or fluctuating temperature (thermal fatigue). Fatigue causes initiation and growth of cracks especially at locations of material discontinuities until the crack size reaches a critical limit such that the asset is no longer able to resist the applied load.

1.1.3 Creep

Creep is continuous plastic deformation that happens when an asset is continuously subjected to a load for a long time. Creep deformation is accelerated at high temperatures. Creep is accompanied by microscopic voids that eventually lead to macroscopic cracks and crack growth.

1.1.4 Corrosion-Fatigue-Creep Interaction

An asset could be subject to combined degradation mechanisms such as corrosion, fatigue and creep. The degradation is accelerated due to the presence of several mechanisms at the same time.

1.2 Maintenance Mechanisms

Maintenance activities such as repair, replacement and alteration may be required as a follow up to inspection. Repair refers to work that is necessary to restore an asset to a state at which it is able to perform its intended function according to the design conditions. Alteration involves a physical change which improves the ability of an asset to perform its intended function. A change in operating conditions such as temperature or pressure is not an alteration and is referred to as re-rating.

Maintenance can be classified as:

1.2.1 Preventive Maintenance

Preventive maintenance consists of maintenance tasks which attempt to pre-empt failure of an asset.

Preventive maintenance can be sub-divided into:

Scheduled Maintenance

Scheduled maintenance is carried out at prescribed intervals of time to ensure that an asset is operating correctly and to avoid any unscheduled downtime into:

Predictive (Condition-Based Maintenance)

Predictive maintenance is based on the use of a physical parameter or characteristic of an asset such as vibration, temperature, pressure, voltage, current, sound, colour or resistance to detect major changes in the performance of the asset. Measurements of the parameter are made either continuously or at regular intervals and the results are compared with initial measurements made when the asset was new. A certain limit, for the amount of acceptable deviation from as new condition, is decided at the beginning of the maintenance cycle. A repair or replacement action is then performed prior to the anticipated time of failure.

1.2.2 Reactive Maintenance (Corrective Maintenance)

Reactive maintenance is carried out after the occurrence of failure in order for an asset to return to its operating condition. This type of maintenance is useful when the cost of the failure consequences is lower than the preventive maintenance cost.

In-service inspection and maintenance could be planned based on one of the following approaches:

The Remaining Life Approach

The remaining life approach is based on calculating the remaining life of an asset according to its tolerance to deterioration, defects or damage and the rate of deterioration. The tolerance to deterioration is determined by assessing fitness-for-service at future times according to the deterioration predicted. The interval between two consecutive inspections equals a fraction of the remaining life. The remaining life is assessed based on analytical fracture mechanics models such as Paris's equation for fatigue crack growth rate or based on statistical methods to estimate the corrosion rate using the history obtained from previous inspections.

Reliability-Based Approach

The inspection and maintenance are planned such that the reliability of the asset remains greater than previously determined target reliability. Reliability centered maintenance (RCM) is a typical example of this approach.

Risk-Based Approach

The inspection and maintenance activities are planned based on maintaining the risk of failure below an acceptable level. RBIM is a typical example of this approach.

Although RBIM reduces risk of failure of a plant subjected to deterioration during its operation, failure may still occur as a result of incorrect design, manufacturing defects, or extreme environmental random events.

Thus, to ensure the integrity of a plant, all the following requirements should be satisfied:

- Proper design procedures should be followed to ensure that all assets are able to withstand the applied loads. Appropriate design approach should be followed and relevant codes and regulations should be consulted.
- A system for quality assurance should be in place to eliminate manufacturing defects.
- An inspection and maintenance plan to ensure the integrity of all assets during operation should be implemented.

RBIM is a team activity. Technical knowledge and experience should include the following:

- Risk Assessment
- Plant potential hazards, damage mechanisms and failure consequences
- Plant safety
- Material and corrosion engineering
- Plant operation and maintenance
- Inspection techniques and inspection history

A number of regulations and standard codes are available to provide the minimum requirements to ensure the integrity of a plant or a specific asset in the three stages of design, manufacturing and operation. Regulations and standard codes which focus on the operation stage aim at developing an effective inspection and maintenance plan for maintaining the integrity of an asset or a plant. Examples of these regulations and standard codes are API 510 (pressure vessels inspection code), API 570 (pipelines inspection code) [3], ASME (risk-based inspection) [6], API 581(risk-based inspection) [4] and PSSR (pressure systems safety regulations) [31].

1.3 Risk-based Inspection and Maintenance (RBIM) Strategies

Many studies which address risk-based inspection and maintenance strategies are available in the open literature. We highlight some of these studies.

Vaurio [35] presented a procedure for optimizing test and maintenance intervals of safety related systems and components. The method is based on minimizing the total plant-level cost under the constraint that the total accident frequency (risk) remains below a set criterion.

Nessim and Stephens [29] presented a risk based methodology that estimates the optimal maintenance interval for an aging hydrocarbon pipeline network. The presented risk based maintenance methodology consists of two main steps: to rank different segments of the pipeline according to priority for maintenance, and to select an optimal set of maintenance actions for the chosen segments.

Balkey, Art and Bosnk [9] developed a methodology, which includes risk based ranking methods, beginning with the use of plant probabilistic risk assessment (PRA), for the determination of risk-significant and less risk-significant components for inspection and the determination of similar populations for pumps and valves for in-service testing. This methodology integrates non-destructive examination data, structural reliability/risk assessment results, PRA results, failure data and expert opinions.

Apeland and Aven [5] developed a risk based maintenance optimization (RBMO) approach. The optimal strategies can be determined by evaluating the relationship between the benefits associated with each maintenance alternative and its cost. The presented approach works in a probabilistic frame using a Bayesian approach.

Kallen [17] developed a probabilistic risk based inspection methodology. This methodology uses a stochastic process to model the corrosion damage mechanism and to develop optimal inspection plans. Cost functions associated with a Gamma process for modeling deterioration are developed.

Misewicz et al. [28] developed a risk based integrity project ranking approach for natural gas and CO_2 pipelines. The approach is based on a benefit cost ratio, defined as the expected risk reduction in dollars per mile over the project useful life, divided by the total project cost. Risk reduction is estimated using a quantitative risk analysis approach. The benefit cost ratio results can be used as a tool to justify the maintenance budget.

Khan and Haddara [20] presented a risk-based maintenance methodology for designing an optimum inspection and maintenance programs. The methodology consists of three parts: risk estimation, risk evaluation and maintenance planning.

Khan and Haddara [21] discussed a comprehensive and quantitative methodology for maintenance planning based on risk. This methodology is developed to obtain an optimum maintenance schedule that minimizes the probability of system failure and its consequences.

Kallen and Noortwijk [18] presented a risk-based inspection (RBI) technique that develops cost and safety optimal inspection plans. Bayesian decision model is used to determine these optimal inspection plans under uncertain deterioration. The presented risk-based inspection technique uses the Gamma stochastic process to model the corrosion damage mechanism and Bayes' theorem to update prior knowledge over the corrosion rate with imperfect wall thickness measurements. A periodic inspection and replacement policy which minimizes the expected average costs per year is found. An example using actual plant data of a pressurized steel vessel is presented.

Krishnasamy, Khan and Haddara [25] developed a risk-based inspection and maintenance methodology for a power generating plant. Applying this methodology reduces risk, increases the reliability of assets, and reduces the cost of maintenance.

Khan and Howard [22] presented a simplified practical approach for the use of statistical tools for inspection planning and integrity assessment. The study is focused on corrosion related material degradation of piping on an offshore production facility.

Khalifa, Khan and Haddara [19] proposed a methodology for the optimal selection of a non-destructive inspection (NDI) technique and its associated in-service inspection interval for welded components subjected to cyclic loading fatigue. An objective function, expressed as the sum of the inspection cost, the repair cost and the risk of failure, is minimized while keeping the probability of failure below an acceptable level.

2 Risk-Based Inspection and Maintenance Approach

Risk-based inspection and maintenance (RBIM) aims at maintaining the integrity of a plant or an asset that is subjected to deterioration. Maintaining the asset integrity guarantees maintaining the risk of failure below an acceptable level.

2.1 Steps of RBIM Planning

An RBIM plan is comprised of the following steps:

2.1.1 Step I. Inspection Sampling

Inspection sampling is a partial inspection of the system where preselected components or areas are inspected.

Layering Separation

Before the sampling process starts, it is required that the plant be classified into sections or groups in order to keep track of the measurements made in each group. This classification is called layering separation. The sections and groups obtained by the classification process are usually referred to as corrosion/damage circuits or loops. A corrosion/damage circuit (loop) is a group of assets in the plant which have the same material, perform under similar operating conditions, subjected to the same damage mechanisms, and exposed to the same environmental conditions. The objective of dividing a plant into corrosion/damage circuits is to reduce the source of variability in the corrosion/damage data within each circuit. This would help to reduce the required sample size when sampling randomly within a corrosion circuit because the sample size is strongly dependent on the standard deviation, σ, of the population.

Sample Size

The appropriate sample size is the number of assets/areas of a corrosion/damage circuit chosen randomly, which will provide an accurate description of the state of the whole circuit. An example of systems in which the application of this technique is beneficial is long unpiggable pipelines. Some pipelines are designed to allow a PIG (piping inspection gauge) to pass through with the fluid to inspect the complete pipeline. These pipelines are called piggable. The inspection of an unpiggable pipeline requires the physical inspection of a large number of points or areas along the pipeline. This may not be practical for very long unpigable pipelines.

The following analysis illustrates how inspection sampling is applied for assets which suffer from either a general metal loss (i.e., general corrosion) or a localized damage (localized metal loss or cracks). The analysis can also be extended to other forms of deterioration.

In case of an asset subjected to general corrosion, the knowledge of the average metal loss and corrosion rate is sufficient to evaluate the time to failure. For evaluating the mean of a population using measurements obtained from a sample, the classical method for determining the sample size may be used. This method is based on keeping the level of error in the mean estimate obtained from the sample (i.e., the deviation between the sample and the population means) within an acceptable limit at a specific confidence level of $(1-\alpha)$. Let "w" denote the maximum acceptable deviation in the mean estimate using the sample from the population mean (for example, if the maximum acceptable deviation in the mean estimate is set at ± 0.1 mm, then $w = 0.1$ mm). w is called the margin of error.

The margin of error, w, can be expressed as half the width of the confidence interval of the mean:

$$w = K(\sigma_{\mu}) = K\frac{\sigma}{\sqrt{n}} \tag{1}$$

where σ_{μ} is the standard deviation of the mean, σ is the standard deviation of the population and K is expressed in terms of the inverse standard normal probability, Φ^{-1}, as follows:

$$K = \phi^{-1}\left(1 - \frac{\alpha}{2}\right) \tag{2}$$

where $(1-\alpha)$ is the confidence level.

This leads to a sample size n given by:

$$n = \left(K\frac{\sigma}{w}\right)^2 \tag{3}$$

When the standard deviation of the population, σ, is unknown and has to be estimated from the data, one can use a t-distribution. In this case, K and σ used in the above equation are replaced with $t_{1-\alpha/2,n-1}$ and S, respectively as follows:

$$n = \left(t_{1-\frac{\alpha}{2},n-1} \cdot \frac{S}{w}\right) \tag{4}$$

$t_{1-\alpha/2,n-1}$ is the critical value at the probability of $(1-\alpha/2)$ of the t-distribution with $(n-1)$ degree of freedom and S is the sample standard deviation. For large sample size n (for example $n > 50$), the t-distribution approaches the standard normal distribution. In this case, the two equations mentioned above give approximately equal sample sizes. The solution for n in the above equation should be obtained by trial and error because $t_{1-\alpha/2,n-1}$ is a function of n.

In case of localized damage, evaluation of the mean value of the damage is not sufficient because the failure is expected when the maximum damage at any location in the asset or system exceeds a critical limit. Thus, the sample should predict the maximum damage over the whole population. It is a practical problem in inspection sampling of localized damage to make a decision regarding the minimum required sample size which represents the population. There is still no clear consensus on this aspect but it can be said that the greater the sample size is the better the reliability of the sample estimate will be [24]. The extreme value statistical method is widely used to predict the maximum damage over all components (inspected and un-inspected) using an inspected sample.

The extreme value distribution is classified into three types (Type I, Type II and Type III) for two cases (maximum values and minimum values). Type I (in case of maximum values) is known as Gumbel's distribution. It is a common practice to use Gumble's distribution to represent the probability distribution of maximum values.

To demonstrate how can one predict the maximum damage over the whole population using the extreme value statistical method, let us consider a system having a total of N components (population of size N) and a sample of size n

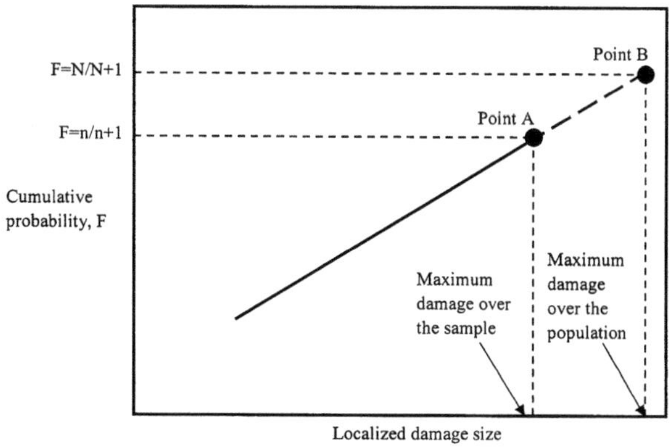

Fig. 1 Extrapolation of the extreme value probability plot

components. The sample cumulative probability distribution can be determined by arranging the data from the lowest to the highest. The cumulative probability is estimated as $i/n+1$ when using the average rank method where i is the rank. The highest value of the sample cumulative probability can be estimated as $F=n/(n+1)$. This maximum value corresponds to the maximum damage over the sample. Similarly, the highest value of the cumulative probability for the whole population of size N can be estimated as $F=N/(N+1)$. The extreme value cumulative probability function is a straight line on extreme value probability plot paper. Thus, the maximum population damage can be predicted by extrapolating the extreme value probability plot linearly from a point A to a point B (Fig. 1). This extrapolation is valid provided that the statistical characteristics of the sample completely represent the statistical characteristics of the whole population.

Structural deterioration is a random process as it is a function of random variables such as crack size, location and orientation. The uncertainty and variability of data obtained by inspection sampling of a population suffering structural deterioration is best modeled by stochastic models.

In engineering applications, it is often the case that one is forced to make maintenance decisions on the basis of limited data. The probability distribution function estimated using inspection sampling of a random variable such as metal loss or crack size is the likelihood probability of this variable. One can use a likelihood probability to update a known prior distribution using Bayesian updating theory. The updated distribution is referred to as a posterior distribution. For example, the probability distribution of metal loss in an asset obtained by inspection sampling can be used to update a known prior distribution. Although the choice of a prior is often subjective, a rational result can be obtained by analyzing historic data from the same or other similar assets. Several techniques are available for estimating a prior distribution using deterioration data. These include frequency diagrams, plotting data using probability graphs, and conducting statistical tests. The parameters of such distributions can be estimated using the methods of least squares and maximum likelihood estimates (MLE), see Thodi, et al. [34].

The following analysis shows how the posterior distribution is obtained for an inspection sample. Let $f''(\theta)$ be the posterior probability distribution of θ. This distribution can be obtained using Bayes' theorem, see Ang and Tang [2]:

$$f''(\theta) = \frac{P(x|\theta)f'(\theta)}{\int_{-\infty}^{\infty} P(x|\theta)f'(\theta)d\theta} \tag{5}$$

where x denotes the observed inspection data. $P(x|\theta)$ is the likelihood or the conditional probability of observing the inspection outcome x assuming a given parameter θ and commonly referred to as the likelihood function of θ. $f'(\theta)$ is the prior probability distribution of θ.

The posterior value, θ'', for the mean of the parameter θ is given by:

$$\theta'' = \int_{-\infty}^{\infty} \theta f''(\theta)d\theta \tag{6}$$

The obtained posterior distribution using Bayesian theory provides a more reliable and adaptive model for handling the uncertainty in inspection sampling data.

2.1.2 Step II. Inspection Planning

An inspection plan involves selection of an inspection technique and inspection schedule.

Selection of an Inspection Technique

Different non-destructive inspection techniques are used for in-service inspection such as ultrasonic, radiographic, eddy current and magnetic techniques. Selection of the best technique to be used depends on the ability of the technique to detect the damage, cost of performing the inspection using this technique, and the risk of failure. This selection could be based on expert subjective judgement or on quantitative basis.

Two main parameters are used to quantify the ability of the NDI techniques to detect a specific flaw. These are the Probability of Detection function (POD) and the Probability of False Calls, (PFC).

The POD function is a measure of the ability of the technique to detect an existing flaw. It is a function of the flaw size, a. The following procedure can be used to estimate the POD function for a specific NDI technique. First, a number of test specimens are chosen. These test specimens may have previously existing flaws. These flaws may have been formed during manufacture or service. Another approach, is to artificially introduce a number of flaws having different sizes. A specific NDI technique is used to detect these flaws. The ratio of the number of flaws detected to the number of the flaws actually existing is calculated. After the completion of the inspection of all samples, the samples are sectioned destructively to verify the presence of the flaws and to measure their sizes. The POD curve obtained is discrete. Each point is representative of a crack class range, and the probability of detection in that class is equal to the ratio of the actual number of detected cracks to the total number of existing cracks in that class.

PFC is defined as the fraction of times that unflawed component will be incorrectly classified as being flawed.

Inspection Scheduling

Inspections are usually scheduled at specified times. Inspection times are chosen based on two factors. The first factor is reducing the risk of failure to an acceptable level. The second factor is the cost of the inspection.

The inspection cost includes the cost for gaining access to the degraded asset, the cost for surface preparation, personnel cost for inspection, the cost associated with technical assistance, the cost of consumables and chemicals, and the logistics cost such as rent, storage and transportation.

2.1.3 Step III. Selection of a Maintenance Activity (Repair, Replacement and/or Alteration)

The condition of the inspected asset will dictate the maintenance action that will be taken. The asset may be repaired, replaced, or left as is depending on its condition. The action taken is based on the acceptable risk of failure of the asset to perform its intended function until the next inspection. The maintenance cost is also a factor when deciding the action to be taken. A balance between the cost of inspection and maintenance will be achieved in the last step (optimization of the RBIM plan).

Repair work includes the access to deteriorated part, surface preparation, cutting and may include a removal and assembling of parts, welding, restoring the protective coating or insulation and testing/re-validation. Thus, in addition to the material cost, the personnel cost of these activities also must be included in the maintenance cost.

2.1.4 Step IV. Risk Assessment

Risk is expressed as the product of the consequences of failure or an undesired event and the probability of its occurrence.

Risk Assessment Approaches

Risk can be assessed using qualitative, quantitative or semi-quantitative approaches.

Qualitative Approach

Qualitative risk assessment relies on subjective judgement. The probability and consequences of failure are expressed in descriptive and relative terms. For example terms like, very unlikely, possible, reasonably probable and probable can be used to describe the likelihood of failure. The consequence of failure may be described as high, moderate, low. The qualitative approach provides an easy and quick method for the assessment of risk. Its disadvantage is that the evaluated risk is subjective and therefore the links to mitigation activities are also subjective.

Quantitative Approach

Quantitative assessment relies on the availability of data. It requires the determination of a specific value for the probability of failure based on probabilistic analysis (not based on subjective judgment) and the determination of a quantitative estimate for the consequences of failure.

An advantage of this approach is that it allows the benefits of the risk mitigation activities to be quantified. This approach requires long time for data recording and analysis.

Semi-Quantitative Approach

In the semi-quantitative risk assessment, numerical values for the probability of failure and the consequences are assumed on the basis of expert judgment using available estimates for similar assets. Then, the tools which are used in the quantitative assessment approach such as the fault tree and event tree can be used to evaluate the system probability of failure using a subjective probability of failure of each asset. Thus, it can be said that the semi-quantitative risk assessment is a combination of the qualitative and quantitative assessments. It does not require as much data as needed for the quantitative approach. In the semi-quantitative risk assessment approach, factors which may affect the probability of failure and the consequences are weighted or given scoring points (for example from 0 to 100) based on expert judgment. The sum of the weights or the scoring points gives an indication of the level of the probability of failure and the consequences of failure. The level of the probability of failure and its consequences can be classified into categories (from the lowest to the highest) and is usually represented in the form of a semi-quantitative risk matrix. The risk level is evaluated as the product of sum of

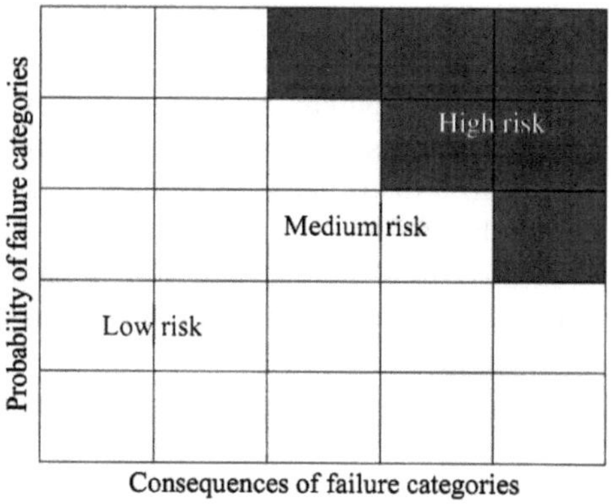

Fig. 2 An example of semi-quantitative risk matrix

weighting factors (or scoring points) of each the probability of failure and consequences. An example of a semi-quantitative risk matrix is shown in Fig. 2.

The inspection plan for each item is then prioritized based on the risk level in the risk matrix.

Failure Modeling

Failure modeling deals with formulating failure models to predict failure occurrences and to evaluate probability of failure. Two approaches may be used to model the failure:

Statistical Failure Modeling Approach

This approach is based on the use of statistical failure data to determine the instantaneous failure rate (hazard function) as a function of time. The probability distribution of time to failure is obtained and the probability of failure is estimated. In this approach, the failure is considered as function of time only without looking at the physical factors or reasons which affect failure. For details of this approach see Ebeling [12].

Physics-of-Failure Modeling Approach

In many applications, the failure of an asset may depend on some variables such as materials properties, damage size, loading, and operating and environment conditions. A more accurate failure model may be one in which these variables are considered. This approach is based on the availability of accurate analytical models to describe the failure of an asset subjected to different degradation mechanisms. This approach requires less data in comparison to the statistical approach. These analytical models are expressed in terms of the variables that affect failure. These variables are in general random. Therefore, this approach can be combined with statistical methods to model the failure. Models obtained using probabilistic fracture mechanics are typical examples of this approach. For example, Paris's equation [30] for modeling the fatigue crack growth rate can be used with a distribution of crack size to estimate the probability of failure of an asset subjected to fatigue.

In physics-of-failure modeling approach, the probability of failure of a deteriorated asset can be expressed as the probability that the limit state function, G, is less than zero.

The limit state function, G, is defined as:

$$G = R - L \tag{7}$$

where R is the resistance or strength and L is the load or stress.

Resistance of an asset subjected to deterioration depends on the material properties in a specified environment, asset dimensions and damage size. Load depends on the operation and environment conditions. For example, the resistance of corroded pipelines is represented in DNV-RP-F101code by capacity pressure, P_{cap}, at which the failure will occur and the load is represented by the internal pressure, P_{cap} of corroded pipeline which can be obtained from DNV-RP-F101code using the following formula:

$$P_{cap} = \frac{2t.SMTS \cdot \left(1 - \frac{d}{t}\right)}{(D - t)\left(1 - \frac{\frac{d}{t}}{Q}\right)} \tag{8}$$

$$Q = \sqrt{1 + 0.31\left(\frac{S}{\sqrt{Dt}}\right)^2} \tag{9}$$

where D is pipe diameter (mm), t is thickness (mm), d is defect depth (mm), S is longitudinal defect length (mm) and $SMTS$ is specified minimum tensile strength (MPa).

Consequences Assessment

The consequences of an asset failure includes loss of commodity due to breakdown, production loss due to shutdown, the legal fees and penalties due to nature damage and the liability cost.

Loss Due to Breakdown

Breakdown costs are the financial losses, which are associated with loosing commodity because of degradation-related failures. This cost depends upon what product is being carried or stored, the rate of leakage and its current market value when the failure occurs. The leak or rupture of a deteriorated asset is the main cause for breakdown

Loss of Production Due to Shutdown

The main factor influencing the cost of failure is the facility's unavailability for production. Inspection and maintenance can be planned, whereas failures may lead to an unplanned, immediate shutdown of assets of the facility. The cost of such a shutdown is highly dependent on the number of days of shutdown, the rate of loss of production and the value of products at the time of failure. The shutdown cost due to deterioration is calculated by combining the unit cost of product, loss of affected production, and maintenance delay time.

Liability Cost

The injuries and deaths caused by a process asset failure have the most severe implications possible. The loss of life or pain of an injury is impossible to quantify, yet, the cost implied due to worker's compensation and corporate liabilities shall be taken into account. Apart from that, safety related system failures have other immediate implications, such as legal fines and penalties of professional negligence. The comprehensive liability cost include medical costs, emergency services, vocational rehabilitation, lost earnings, administrative costs, legal consulting fees, pain and lost quality of life.

Loss Due to Nature Damage

The size of penalty that the company will incur as a result of damaging the environment is difficult to estimate, because costs increase with the scope of failure. The failure modes developed for each failure could be graduated to more complex system failures leading to significant environmental damages. The cost due to damage to natural resources is also difficult to estimate. Still, approximate assessments considering the quantity of release and unit rate are quite possible [13]. The total cost of environmental damage due to deterioration-related-failure comprises of unit cost of nature damage and the total quantity released. The total quantity released from a facility depends on the rate of release and the duration of release.

The release could have the potential for one or a combination of the following events:

Flammable releases: when flammable releases meet with a source of ignition a fire will result at a close range but an explosion may reach over a large range from the centre of the release.

Steam and hot gas releases: can result in very severe injury or burns to people within range.

Toxic releases: the dispersion of toxic releases may extend over large distances from the site. Employees, the general public and the environment may be affected.

High pressure gas release: high pressure gas release has the potential to cause physical injury to personnel in the vicinity and cause structural damage to surrounding assets.

Uncertainties Handling in Risk Assessment

Many tools are used to assess the risk. These include failure mode effect and criticality analysis (FMECA), hazard and operability (HAZOP), layers of protection analysis (LOPA), Markov analysis, fault tree analysis (FTA), Event tree analysis (ETA) and probabilistic fracture mechanics (PFM).

Event Tree Analysis (ETA) and Fault tree Analysis (FTA) are two methods for quantitative risk assessment that can be used to develop a logical relationship among the events leading to a failure and the estimate of risk associated with the failure. ETA is a technique used to describe the consequences of an event (initiating event) and estimate the likelihoods (frequency) of possible outcomes of the event. FTA represents basic causes of occurrence of an unwanted event and estimates the likelihood (probability) as well as the contribution of different causes leading to the unwanted event.

Event and fault trees help to conduct the quantitative risk assessment based on two major assumptions. Firstly, the likelihood of events or basic-events is assumed to be exact and precisely known, which is not very often true due to inherent uncertainties in data collection and the definition of the relationships between events or basic-events [14, 33]. Moreover, because of variant failure modes, design faults, poor understanding of failure mechanisms, as well as the vagueness of system phenomena, it is often difficult to predict the acquired probability of basic-events/events precisely. Secondly, the events or basic-events in an event tree or fault tree are assumed to be independent, which is often an inaccurate assumption. These two assumptions lead to misrepresentation of the process system actual behaviours and impart two different types of the uncertainties, namely data uncertainty and dependency uncertainty.

Fault and event trees can be analyzed either deterministically or probabilistically. The deterministic approach uses the crisp probability of events (or basic-events) and determines the probability of the top-event and the frequency of outcome events in the fault and event trees, respectively. The probabilistic approach treats the crisp probability as a random variable and describes uncertainty using probability distributions [15, 37]. Traditionally, the probabilistic approach uses Monte Carlo Simulation (MCS) to address the random uncertainty in the inputs (i.e., probability of basic-events or events) and propagates the uncertainty for the outputs. The probability distributions for the inputs can be derived from historical information, but are often rare especially when the system is comprised of thousands of assets.

FTA and ETA require probability data of events (or basic-events) as inputs to conduct a comprehensive quantitative risk assessment for a system. Since most of the time the crisp data as well as probability distributions are rarely available for all events and basic-events, expert judgment/knowledge are often employed as an alternative to the objective data. Two types of uncertainties, namely aleatory and epistemic uncertainties, are usually addressed while using the expert's knowledge in quantitative risk assessment [7, 14]. Aleatory uncertainty is a natural variation, randomness or heterogeneity of a physical system. It can be well described using probabilistic methods if enough experimental data are available to support the analysis [1]. Epistemic uncertainty means ambiguity and vagueness, ignorance, knowledge deficiency, or imprecision in system behaviours.

In quantitative risk assessment, it is important to characterize, represent, and propagate the uncertainty accurately in order to get a reliable analysis. However, when the input probability distributions are 'reasonably known', Monte Carlo simulation (MCS) can be used to estimate and propagate the uncertainties, especially two dimensional MCS which can effectively deal with both aleatory and epistemic uncertainties [8]. If knowledge is limited for definition of the probability distributions, probabilistic approaches might not be the best choice to handle the uncertainty in quantitative risk assessment [11]. In addition, the independence assumption of events (or basic-events) might be convenient to simplify the FTA or ETA, however it is not always true for all cases [16]. This assumption in fact adds another kind of uncertainty, i.e., the dependency uncertainty. Vesely et al. [36] show several cases of FTA where the assumption that the basic-events are independent is not valid.

Fuzzy set and evidence theories have recently been used in many engineering applications where expert knowledge is employed as an alternative to crisp data or probability distributions. Fuzzy set theory is used to address the subjectivity in expert judgment. Whereas, evidence theory is more promptly employed in handling the uncertainty which arises due to ignorance, conflict and incomplete information, in addition to the uncertainty in the dependency among the basic-events in the FTA. Li [27] proposed a fuzzy approach based dependency factor to address the dependencies in performing risk assessment. The probabilities of events (or basic-events) and their dependency coefficients which are derived through expert knowledge can be treated as fuzzy numbers. The fuzzy numbers in fuzzy set theory describe linguistic and subjective uncertainty while basic probability assignments in evidence theory are used to handle ignorance, incompleteness and inconsistency in expert knowledge. A generic framework is shown in Fig. 3 illustrating the use of fuzzy set theory and evidence theory to handle two different kinds of uncertainties in FTA and ETA.

The probability of events (or basic-events) can be defined linguistically and described using triangular fuzzy number (TFN). The interdependence of events (or basic-events) is defined linearly using a dependency coefficient (Cd) that can also be described using a TFN. Fuzzy probability and dependency coefficients are used

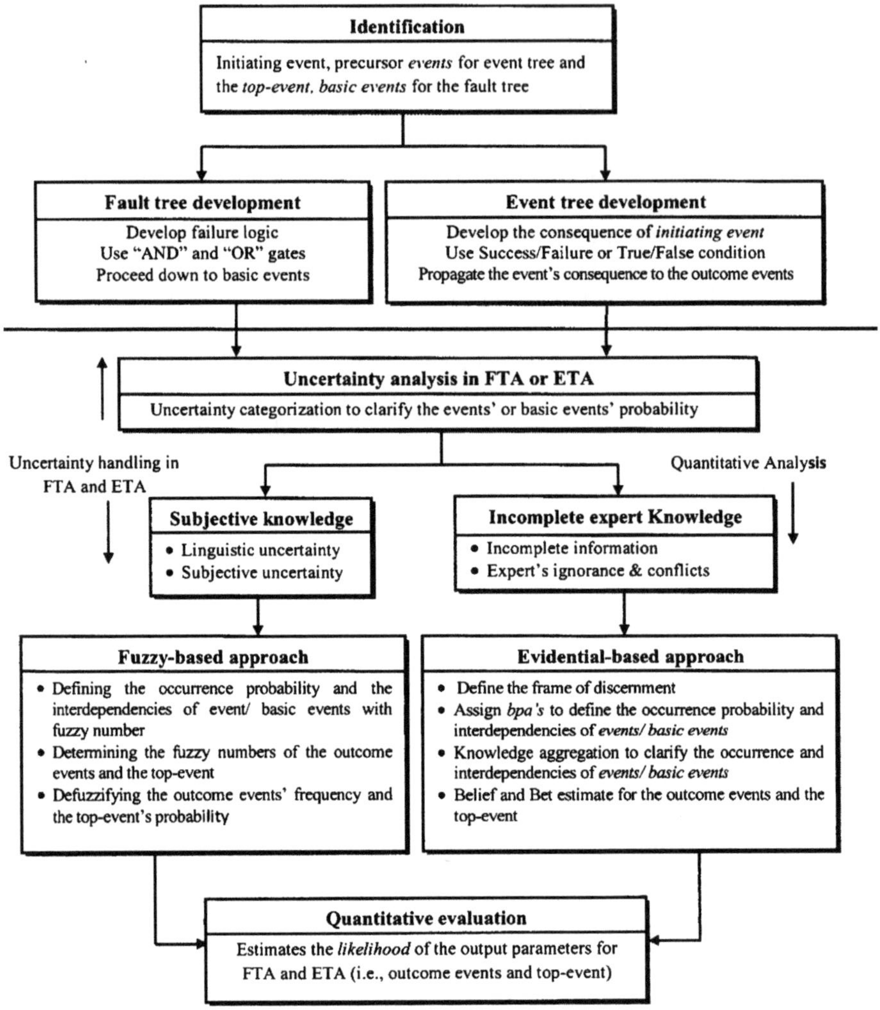

Fig. 3 Framework for FTA and ETA under uncertainty

to determine the probability of top-event and the frequency of outcome events in fuzzy terms.

2.1.5 Step V. Optimization of the RBIM Plan

The optimization of an inspection and maintenance plan aims at minimizing the total lifetime cost without compromising the plant integrity. An objective function can be formulated as sum of the inspection cost, maintenance (repair/replacement/alteration) cost and risk of failure (expressed as cost).

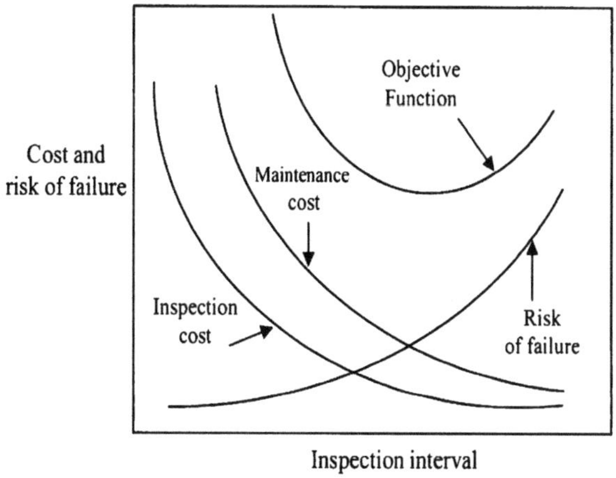

Fig. 4 Cost (inspection and maintenance), risk of failure and objective function versus different inspection intervals

Cost of inspection and cost of maintenance increase as the inspection interval decreases because the shorter the inspection interval the greater the expected number of inspections during the lifetime will be. This leads to high cost of inspection and high cost of maintenance. The longer the inspection interval is the higher the risk of failure will be.

The inspection cost, the maintenance cost and the risk of failure are estimated for different inspection intervals. The objective function is then estimated for the different inspection intervals. As inspection cost and maintenance cost are decreasing with increasing the inspection interval while the risk of failure is increasing with increasing the inspection interval, therefore, the obtained objective function should be having valley shaped (Fig. 4).

The optimum inspection interval can be obtained at the minimum value of the valley shaped objective function. The solution of the RBIM optimization problem is subject to a constraint that the risk of failure does not exceed an acceptable level. Other constraints could be applied such as constraints dictated by the maintenance budget or the maximum inspection interval allowed by regulations or law.

The solution of the optimization problem can be repeated for different maintenance activities (repair, replacement and/or alteration) and the minimum values of the objective function can be compared to obtain the optimum inspection and maintenance plan

The future annual inspection and maintenance costs tend to increase with time. An escalation rate may be used to predict the costs of labor and materials in future at different points of time.

All costs in the objective function at different points of time are to be discounted to a net present value (NPV) when comparing different inspection and maintenance plans.

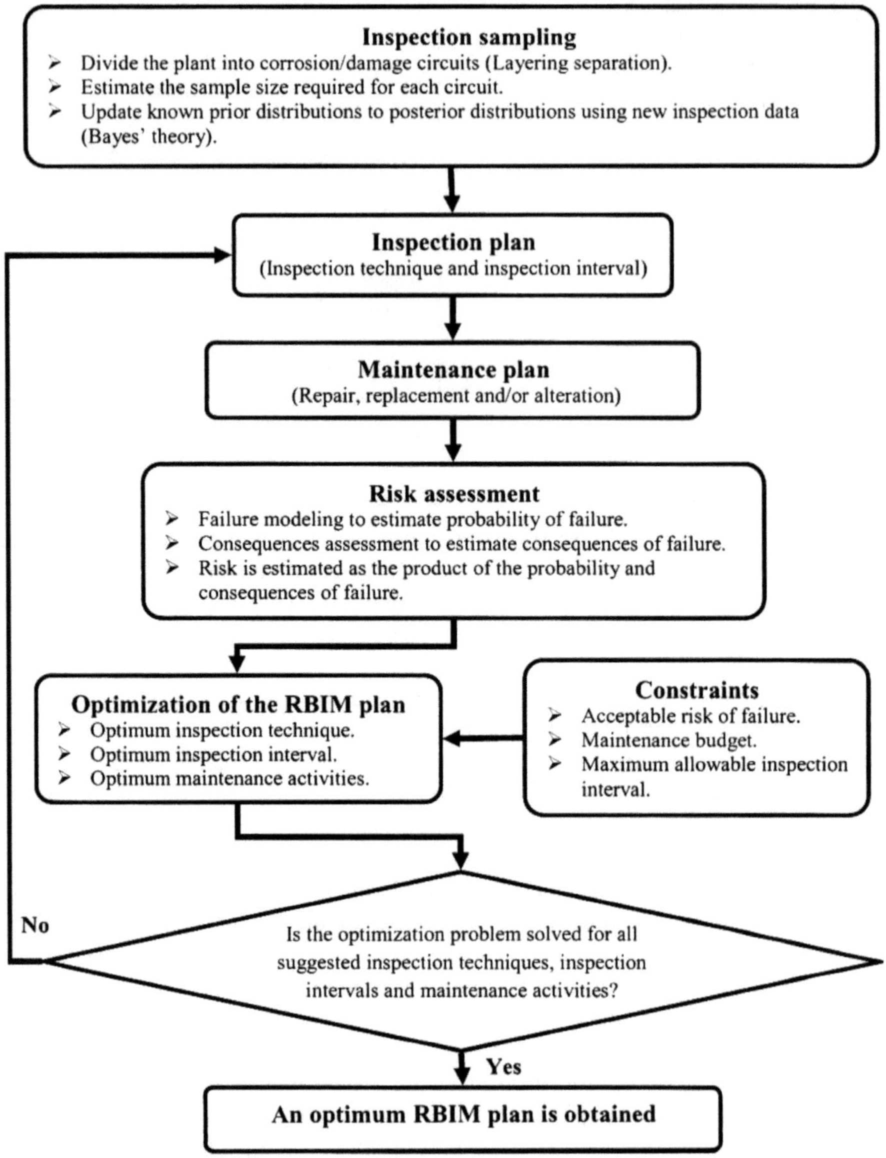

Fig. 5 RBIM framework for a power plant

3 RBIM Framework for a Power Plant

The steps of the RBIM planning for a power plant are shown in the framework presented in Fig. 5.

4 Case Studies

Two case studies are presented, the first for an asset and the second for a complete unit in a power plant. A welding joint in a pipeline is selected for the first case study as the welded joints in power plants could be critical due to welding flaws and stress concentration. The second case study is for a complete power-generating unit. Physics-of-failure modeling approach is used to model the failure in the first case study while the statistical failure modeling is used in the second case study.

4.1 Case Study 1: A Welding Joint Subjected to Fatigue

A welding joint located in 22″ *OD* (thickness = 12.9 mm) pipeline subjected to fatigue caused by vibrations is considered in this case study.

The distribution of the stress range ($\Delta\sigma$) due to variations in stress is modeled by Weibull distribution with scale parameter of 13.4 MPa and shape parameter of 1.5. The average number of stress cycles is taken 2.4×10^6 cycle/year.

All existing welding flaws are conservatively assumed cracks with initial size a_o. The initial size, a_o, is modeled as a lognormally distributed random variable with a mean value of 0.97 mm and a standard deviation of 0.504 mm. The critical crack size at which leakage will occur, a_{cr} is considered to be constant at 5 mm for this case study. The fatigue crack growth parameter, C, is modeled as a lognormal variable with a mean value of 6.06×10^{-13} assuming units of millimetres for crack size and MPa.mm$^{1/2}$ for fracture toughness and a standard deviation of 1.58×10^{-13} mm. The fatigue crack growth parameter, m, is modeled as a normally distributed random variable with a mean value of 3 and a standard deviation of 0.14. The repair policy dedicates the repair for any detected crack size (i.e., the crack size which should be repaired, a_r, is assumed 0).

Maximum acceptable probability of failure to detect a growing crack before reaching the critical size is taken as 10^{-3}. This is the constraint of the RBIM optimization problem in this case study.

A methodology for RBIM optimal planning for welded assets subjected to fatigue cracking proposed by Khalifa et al. [19] is used in this case study as follows:

The objective function, *OF*, is given by:

$$OF = E[C_I] + E[C_R] + E[C_F] \tag{10}$$

where $E[C_I]$ is the expected cost of inspections over the lifetime, $E[C_R]$ is the expected cost of repairs over the lifetime and $E[C_F]$ is the expected cost of failure.

The optimization problem is defined as minimizing the value of the objective function subject to a safety constraint that probability of failure to detect a growing crack before reaching the critical size, P_F, does not exceed the predefined acceptable level.

$E[C_I]$ can be estimated as:

$$E[C_I] = E[n]K_I \tag{11}$$

where K_I is cost of one inspection and $E[n]$ is the expected number of inspections over the lifetime which can be estimated as follows:

$$E[n] = \frac{Mean\_time\_of\_failure}{Inspection\_Interval} = \frac{t_{cr.mean}}{t_{\text{int}}} \tag{12}$$

The critical time to failure is estimated using the well known Paris's equation [30] which relates crack growth to the number of stress cycles. This is the physics-of-failure model as follows:

$$\frac{da}{dN} = C(\Delta K)^m \tag{13}$$

where a is the crack size, N is the number of stress cycles, C and m are material parameters for fatigue crack growth and Δk is the stress intensity range factor which, in general, can be calculated as follows:

$$\Delta K = F(a)\Delta\sigma\sqrt{\pi a} \tag{14}$$

where $\Delta\sigma$ is the applied stress range and $F(a)$ is the geometry function. The geometry function $F(a)$ is taken unity for simplicity in this case study.

An integral form of Paris Law is given by:

$$N_{cr} = \int_{a_o}^{a_{cr}} \frac{da}{C[\Delta\sigma.F(a)\sqrt{\pi a}]^m} \tag{15}$$

where N_{cr} is the critical number of cycles to failure, a_o is the initial crack size and a_{cr} is the critical crack size at which the failure is expected.

The critical time to failure, t_{cr} is obtained as multiplication of the critical number of cycles and frequency of loading (e.g. 500,000 cycle/year) for combinations of random variables (a_o, C, m and $\Delta\sigma$). These random variables are generated using Monte Carlo simulation method. The mean time to failure,$t_{cr,mean}$, is estimated as the average of critical time to failure estimated from Paris law for each combination of (a_o, C, m and $\Delta\sigma$).

The expected cost of failure, $E[C_F]$ is estimated as follows:

$$E[C_f] = P_f K_f \tag{16}$$

$$P_f = \frac{\sum_{j=i}^{N_1}\left[\prod_{i=1}^{n_j}[(1-POD)H_i]\right] = N_2}{N} \tag{17}$$

Table 1 Probability of detection obtained by Berens and Hovey [10]

$POD(a) = \frac{\exp(A+B.\ln a)}{1+\exp(A+B.\ln a)}$		
NDI technique	A	B
UI	−0.119	2.986
MI	0.466	0.604

where P_F is probability of failure and K_F is cost of failure consequences, N is total number of simulations, N_1 is number of simulations at which the first inspection time is less than or equal the critical time to failure and N_2 is the number of simulations at which the first inspection time is more than the critical time, n_j is number of inspections before reaching the critical limit in simulation number j, POD_i is probability of detection function = $POD(a_i)$ where a_i is the crack size at time of the ith inspection. H_i is probability of presence a crack at time of the ith inspection which can be calculated as follows:

$$H_i = \int_0^{a_i} f_i(a)da \tag{18}$$

a_i and $f_i(a)$ are crack size and probability density function of the crack size, a, at time of the *ith* inspection, respectively.

The expected cost of repairs, $E[C_R]$, is estimated as follows (see Khalifa et al. [19]):

$$E[C_R] = K_R \left[\frac{\sum_{j=1}^{N_1}\sum_{i=1}^{n_j} [PFC\,(1-H_i)]}{N} + \frac{\sum_{j=1}^{N_i}\sum_{i=1}^{n_j} [POD_i\,H_i\,(1-A_i)]}{N} \right] \tag{19}$$

where K_R is the cost of one repair.

$$A_i = \begin{cases} 1, & a_i \le a_r \\ 0, & a_i > a \end{cases} \tag{20}$$

A_i is probability of accepting a crack of size a_i and a_r is the limit of crack size at which any detected crack larger than this limit should be repaired.

Two NDI techniques are considered in this example, Ultrasonic inspection (UI) and Magnetic Inspection (MI). POD functions for UI and MI are assumed to be given as shown in Table 1 Berens and Hovey [10].

The relative costs of inspection (K_I), cost of repair, K_R, and cost of failure, K_F, are taken as:

$$K_{I.MI} : K_{I.UI} : K_R : K_F = 1.2 : 1.5 : 10 : 20000 \tag{21}$$

PFC for each inspection technique is assumed to be 1.4% for UI and 5% for MI.

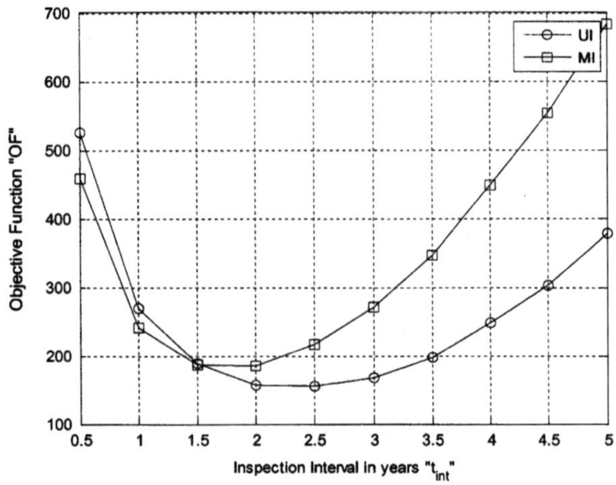

Fig. 6 Objective function versus different inspection intervals

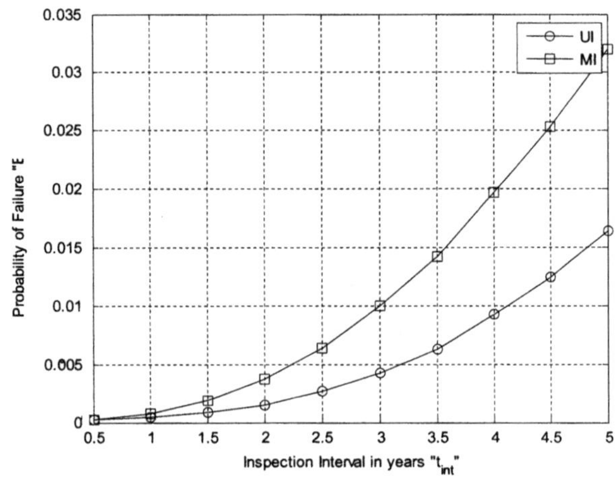

Fig. 7 Probability of failure versus different inspection intervals

By applying the proposed methodology by Khalifa et al. [19], the obtained objective function is shown in Fig. 6 for both UI and MI.

From Fig. 7, the maximum acceptable inspection interval which keeps probability of failure not exceeding 10^{-3} (safety constraint) is 1.5 year for UI and 1 year for MI.

From Fig. 6, the minimum value of the objective function is 155.35 (located at inspection interval 2.5 years) for UI and 184.6307 (located at inspection interval 2 years) for MI. By comparing the minimum value of the objective function of the two inspection techniques, it is preferable to use UI technique with inspection interval 2.5 year, but for the safety constraint, the inspection interval should not exceed 1.5 year for UI and 1 year for MI. By comparing the value of the objective function at inspection interval of 1.5 years for UI which is 188.30 and at 1 years for MI which is 241.55, leads to *the optimum selection between the two techniques (UI and MI) is UI with inspection interval of 1.5 years*. This selection shall ensure the minimum possible value of the objective function taking into consideration the safety constraint, probability of failure is less than 10^{-3}.

Figure 7 shows probability of failure versus different inspection intervals.

4.2 Case Study 2: A Power-Generating Unit

This case study was presented by Krishnasamy, Khan and Haddara [25]. The data used in this case study was obtained from Unit 3 of Holyrood power-generating plant located in Newfoundland, Canada. It has a rated capacity of 150 MW. Unit 3 is classified into major subsystems based on the operational characteristics. A subsystem is comprised of different assets or devices such as pumps, feed water heaters, valves and soot blowers. Fig. 8 shows the hierarchy of systems and subsystems of Unit 3 (in terms of their logical classification). Failure data for the basic assets were obtained from the power plant records. Both the Weibull and the exponential distributions were used to model time to failure of each asset.

Parameters of time to failure Wiebull distribution (β and θ) and exponential (λ) of various assets of Unit 3 sub-systems are estimated.

Fault trees were constructed for the different plant systems. An example is given in Fig. 9, which depicts the fault tree for the event "failed to generate and supply power". Each basic event of this fault tree (a total of 13 basic events as shown in Fig. 8) was subsequently extended in one or more fault trees and analyzed. Using the results of this analysis, one can determine the probability of occurrence of these basic events. A software package 'PROFAT' was used to analyze these fault trees, see Khan and Abbasi [23].

In arriving at the top event probability using a fault tree, failure probabilities for the basic events were mostly determined using failure data obtained from the physical plant. However, some data were lacking, and for these assets, failure rates were estimated either from reliability data banks [26, 32] or from operating experience of plant personnel (semi-quantitative assessment).

Consequence analysis involves the estimation of maintenance cost and the production loss costs during the expected down time if failure occurs. The down time includes supply delay, diagnosis time, replacement/repair time and revalidation time. Maintenance cost is comprised of labor and parts costs.

Risk of failure over 20 years is calculated by multiplying the probability and the consequence of failure as shown in Table 2.

An acceptable risk criterion was determined based on the yearly maintenance expenditure of Unit 3 (found from records as $2,000,000 per year). The estimated risk for each individual subsystem was compared against the acceptable risk criterion. Subsystems whose estimated risk exceeded the acceptance criteria were identified. These are the units whose maintenance plan had to be modified in order to lower their risk. To facilitate this comparison, a risk index was calculated.

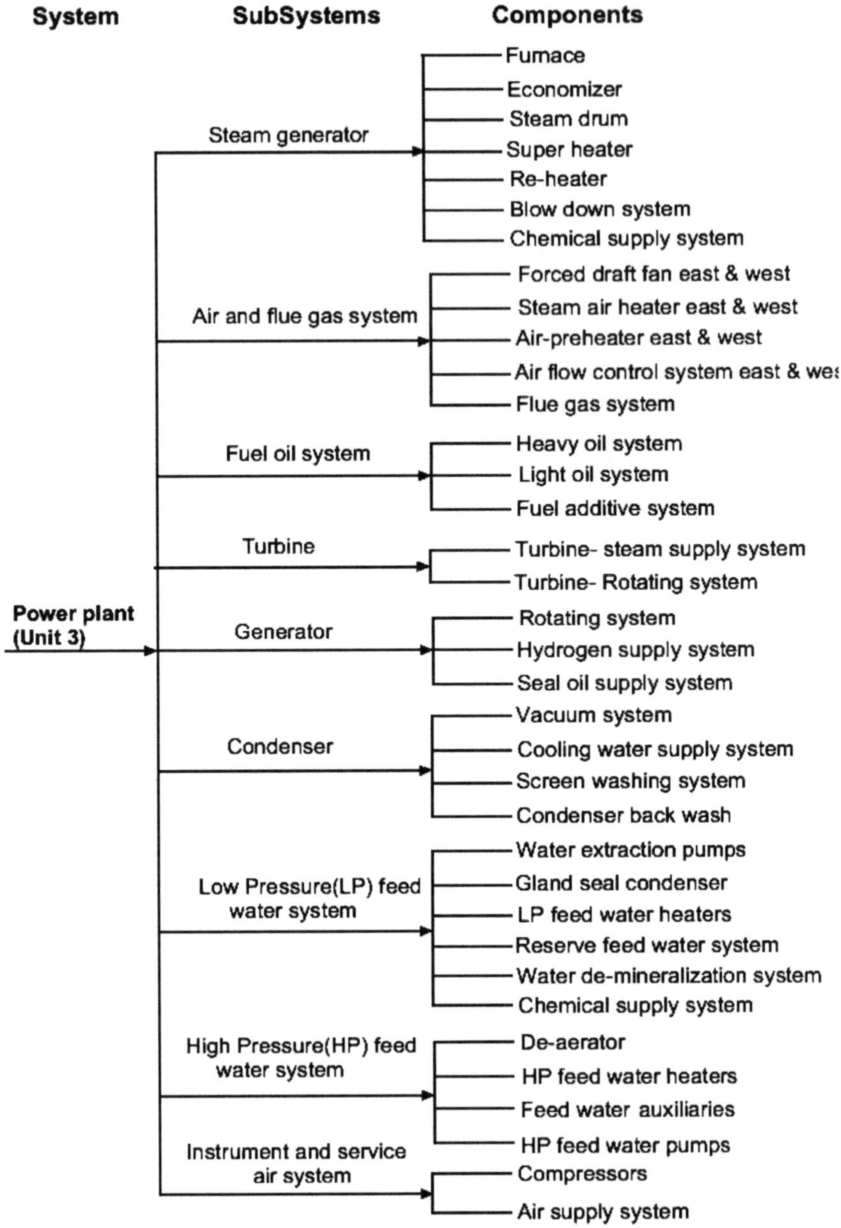

Fig. 8 Classification of unit 3

The risk index is the actual risk divided by the acceptable risk. Thus, any subsystem whose risk index is greater than 1.0 is considered.

Three subsystems were found to violate the risk criterion: the steam generator, air and flue gas system, and the high pressure feed water. A new maintenance

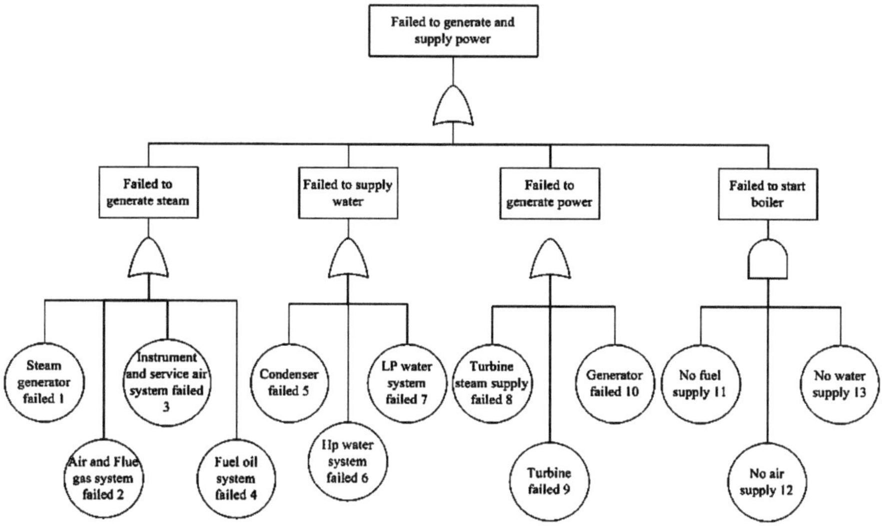

Fig. 9 Fault tree for a failure scenario in unit 3

Table 2 Risk analysis results

Rank	Major system	Consequence in millions	Probability of failure over 20 years	Risk ($) over 20 years	Risk index
1	Steam generator	3,678,481	0.9989	3,674,435	1.837
2	High pressure feed water system	2,478,842	0.9999	2,478,594	1.239
3	Air and flue gas system	2,102,023	0.9914	2,083,946	1.042
4	Generator	1,634,060	0.9780	1,598,111	0.799
5	Turbine-steam supply	1,110,574	0.9999	1,110,463	0.555
6	Fuel oil system	1,110,574	0.9866	1,095,692	0.548
7	Condenser	874,745	0.9939	869,409	0.403
8	Turbine rotating system	302,053	0.9999	302,023	0.151
9	Low pressure feed water system	286,584	0.9995	286,441	0.143
10	Instrument and service air system	25,249	0.9650	24,365	0.012

schedule had to be developed for these three subsystems. To find out which assets contribute more the high-risk levels of these subsystems, a study of the assets of the subsystems was carried out. The assets were divided into three categories, high risk (risk index value greater than 0.8), medium risk (risk index value between 0.4 and 0.8), and low risk (risk index value less than 0.6). The results showed that only one asset (air preheater east) is at high risk while most of the sub-systems/components are at medium and low risk. Tables 3 and 4 show the sub-systems/components which are at medium and low risk respectively.

Table 3 Sub-systems/components of Unit 3 which are at medium risk

Sub-systems/components	Risk value ($)	Risk index
Forced draft fan east	1,444,656	0.7278
Forced draft fan west	1,333,840	0.6669
Heavy oil system	1,109,352	0.5547
Re-heater	1,107,242	0.5536
Super heater	1,102,245	0.5511
Furnace	918,590	0.4593

Table 4 Assets of unit 3 which are at low risk

Sub-systems/components	Risk value ($)	Risk index
Air preheater west	270,734	0.1354
Flue gas system	123,272	0.0616
Air flow control system west and east	108,783	0.0544
Air flow control system east	108,783	0.0544
Steam air heater west and east	108,658	0.0543
Steam air heater west and east	108,658	0.0543
Economizer	79,781	0.0399
Steam drum	73,312	0.0367
Blow down system	32,472	0.0162
Vacuum system	19,827	0.0099
Water extraction	15,374	0.0077
Cooling water supply system	12,827	0.0064
Screen wash system	12,618	0.0063
Light oil system	11,568	0.0058
Fuel additive system	18,350	0.0092
Low pressure heater #1	8,372	0.0042
Low pressure heater #2	8,290	0.0041
Reserve feed water system	7,192	0.0036
Gland seal condenser	7,165	0.0036
Water demineralization system	6,894	0.0034
Condenser back wash	2,982	0.0015
Chemical supply system	2,338	0.0016

The strategy that is adopted to lower the risk to meet the acceptable criterion was to reduce the probability of failure. A probability of failure for the top event was determined such that the resulting risk would be acceptable. A reverse fault tree analysis was used to obtain the probability of failure of the basic events, which would produce a probability of failure for the top event equal to modified value obtained by meeting the risk criterion. The reverse fault tree analysis involves top to bottom analysis approach. Here the probability of occurrence of the top event is fixed and the fault tree is simulated to calculate probability of failure of basic events. The simulation is carried out for many different scenarios, the scenario giving most realistic failure probabilities are accepted. The new probabilities of

Table 5 Risk reduction results

Subsystem	Initial risk factor ($)	Target reduction in probability of failure	Achieved risk reduction in dollars
Steam generator	3,674,434	0.54	1,984,194
Air and flue gas system	2,083,945	0.85	1,771,353
HP feed water system	2,478,594	0.80	1,982,875

Table 6 Maintenance intervals for the super heater

Assets	Maintenance interval
Secondary super heater (SS)	1 year
Primary super heater (PS)	1 year
SS inlet and outlet headers	10 years
PS inlet and outlet headers	10 years
Safety valves	3 months
Temperature indicating transmitters	6 months
Steam and control system	3 months
Attemperator	1 year
Control valve	3 months
Pressure indicating transmitters	6 months
Boiler control	3 months
Combustion control	3 months
Fuel oil management and control	1 year
Spray nozzle	1 year
Globe valve	6 months
By pass valve	6 months

failure of the basic events were used to calculate the corresponding maintenance interval using the probabilistic failure model developed earlier.

The critical subsystems based on risk are identified. Three sub-systems were found to have unacceptable initial risks. These are the steam generator, the high pressure (HP) feed water system, and the air and flue gas system. These three sub-systems are responsible for about 62% of the overall risk of Unit 3. Reducing the individual risk of each of these assets will result in an over all reduction in the risk of the unit. Table 5 shows the risk reduction in dollars for these three sub-systems.

The maintenance intervals are estimated for all assets. In deciding the maintenance interval, the assets that would be maintained at the same time are grouped and assigned the minimum length of the maintenance interval for the whole group. This means that some assets will be over maintained. However, the resulting savings in terms of reducing the down time justify this policy. An example of the estimated maintenance intervals is given for the super heater in Table 6.

References

1. Agarwal H, Renaud EJ, Preston LE et al (2004) Uncertainty quantification using evidence theory in multidisciplinary design optimization. Reliab Eng Syst Saf 85:281–294
2. Ang A, Tang W (2007) Probability concepts in engineering. Wiley, New York
3. API 570 (1998) Piping inspection code. inspection, repair, alteration and rerating of in-service piping systems. API publishing Services, Washington
4. API 581 (2000) Risk based inspection resource document. API publishing Services, Washington
5. Apeland S, Aven T (1999) Risk based maintenance optimization: foundational issues. Reliab Eng Syst Saf 67:285–292
6. ASME (1999) Risk Based Inspection (RBI): A risk-based approach to planned plant inspection. Health and Safety Executive—Hazardous Installations Division, CC/TECH/SAFETY/8
7. Ayyub B, Klir JG (2006) Uncertainty modeling and analysis in engineering and the sciences. Chapman and Hall/CRC, New York
8. Baraldi P, Zio E (2008) A combined monte carlo and possibilistic approach to uncertainty propagation in event tree analysis. Risk Anal 28(4):1–17
9. Balkey KP, Art RJ, Bosnk RJ (1998) ASME Risk-Based in service inspection and testing: an outlook to the future. Risk Anal 18(4):407–421
10. Berens AP, Hovey PW (1981) Evaluation of NDE reliability characterization. Wright-Patterson Air Force Base, Dayton
11. Druschel RB, Ozbek M, Pinder G (2006) Application of Dempster-Shafer theory to hydraulic conductivity. In: Paper of the CMWR—XVI, Conference Program. Copenhagen, 18–22 June 2006
12. Ebeling EC (1997) An introduction to reliability and maintainability engineering. McGRAW-HILL, New York
13. Etkin DS (2000) Worldwide analysis of marine oil spill cleanup cost factors. In: Paper of the Arctic and Marine Oil spill Program Technical Seminar. Environment Canada, June 2000
14. Ferdous R, Khan F, Sadiq R et al (2009) Methodology for computer aided fuzzy fault tree analysis. Process Saf Environ Protect, IChemE 87:217–226
15. Ferdous R, Khan F, Sadiq R et al (2009) Handling data uncertainties in event tree analysis. Process Saf Environ Protect, IChemE 87:283–292
16. Ferson S, Hajagos J, Berleant D et al. (2004) Dependence in Dempster-Shafer theory and probability bounds analysis. US: Sandia National Laboratories, New York
17. Kallen MJ (2002) Risk-based inspection in the process and refining industry. Msc Thesis, University of Delft, Delft
18. Kallen MJ, Noortwijk JM (2004) Optimal inspection and replacement decisions for multiple failure modes. In: Paper of the Proceedings of the 7th International Conference on Probabilistic Safety Assessment and Management. Berlin, 14–18 June 2004
19. Khalifa M, Khan F, Haddara M (2009) Optimal selection of non-destructive inspection technique for welded components. Br J Nondestr Test 51(4):192–200
20. Khan FI, Haddara M (2003) Risk-based maintenance (RBM): a quantitative approach for maintenance/inspection scheduling and planning. J Loss Prev Process Indust 16:561–573
21. Khan FI, Haddara M (2004) Risk-based maintenance (RBM): a new approach for process plant inspection and maintenance. Process Saf Progr 23(4):252–265
22. Khan FI, Howard R (2007) Statistical approach to inspection planning and integrity assessment. Nondestr Test Cond Monit 49(1):26–36
23. Khan FI, Abbasi SA (2000) Analytical simulation and PROFAT II: A new methodology and a computer automated tool for fault tree analysis in chemical process industries. J Hazard Mater 75:1–27
24. Kowaka M, Tsuge H, Akashi M et al (1994) Introduction to life prediction of industrial plant materials: Application of extreme value statistical method for corrosion analysis. Allerton Press, New York

25. Krishnasamy L, Khan F, Haddara M (2005) Development of a risk-based maintenance (RBM) strategy for a power-generating plant. J Loss Prev Process Indust 18(2):69–81
26. Lees FP (1996) Loss prevention in the process industries. Butterworths, London
27. Li H (2007) Hierarchical risk assessment of water supply systems. PhD Thesis, Loughborough University, p 235
28. Misewicz D, Smith AC, Nessim M et al. (2002) Risk based integrity project ranking. In: Paper of the 4th International Pipeline Conference (IPC2002), Calgary, Sept. 29 to 3 Oct 2002
29. Nessim MA, Stephens MJ, Zimmerman TJ (1995) RiskRisk based maintenance planning for offshore pipelines. Proceedings of the Annual Offshore Technology Conference 2:791–800
30. Paris PC, Endogan F (1963) Critical analysis of crack propagation laws. Basic Eng 85:528–534
31. Pressure Systems Safety Regulations, PSSR (2000) Safety of pressure systems. Approved Code of Practice (ACoP) SI-2000-128. The Health and Safety Executive (HSE)
32. AC R (2002) Non-electric components reliability data. Center for Reliability Assessment, New York
33. Sadiq R, Saint-Martin E, Kleiner Y (2008) Predicting risk of water quality failures in distribution networks under uncertainties using fault-tree analysis. Urban Water J 5(4):287–304
34. Thodi PN, Khan FI, Haddara MR (2009) The selection of corrosion prior distributions for risk based integrity modeling. Stoch Environ Res Risk Assess 23:793–809
35. Vaurio JK (1995) Optimization of test and maintenance intervals based on risk and cost. Reliab Eng Syst Saf 49:23–36
36. Vesely WE, Goldberg FF, Roberts NH et al. (1981) Fault treeFault tree handbook. U.S. Nuclear Regulatory Commission, NUREG-0492, Washington
37. Wilcox CR, Ayyub MB (2003) Uncertainty modeling of data and uncertainty propagation for risk studies. IEEE Proceedings on Uncertainty Modeling and Analysis, pp 184–191

Index

G. F. M. de Souza (ed.), *Thermal Power Plant Performance Analysis*, Springer Series in Reliability Engineering, DOI: 10.1007/978-1-4471-2309-5,